Concepts & Images

Design Science Collection

Series Editor
Arthur L. Loeb
Department of Visual and Environmental Studies
Carpenter Center for the Visual Arts
Harvard University

Amy C. Edmondson	*A Fuller Explanation: The Synergetic Geometry of R. Buckminster Fuller, 1987*
Marjorie Senechal and George Fleck (Eds.)	*Shaping Space: A polyhedral Approach, 1988*
Judith Wechsler (Ed.)	*On Aesthetics in Science, 1988*
Lois Swirnoff	*Dimensional Color, 1989*
Arthur L. Loeb	*Space Structures: Their Harmony and Counterpoint, 1991*
Arthur L. Loeb	*Concepts and Images*

Design Science Collection
A Pro Scientia Viva Title

Arthur L. Loeb

Concepts & Images

Visual Mathematics

with 160 Illustrations

Birkhäuser
Boston · Basel · Berlin

Arthur L. Loeb
Department of Visual and
Environmental Studies
Carpenter Center for the Visual Arts
Harvard University
Cambridge, MA 02138

Library of Congress Cataloging-in-Publication Data

Loeb, Arthur L. (Arthur Lee)
Concepts and images : Visual mathematics / Arthur Loeb.
p. cm. -- (Design science collection)
"A pro scientia viva title."
Includes bibliographical references and index.
ISBN 0-8176-3620-X (alk. paper). -- ISBN 3-7643-3620-X (alk. paper)
1. Mathematics. 2. Geometry. 3. Design. I. Title. II. Series.
QA36.L64 1992 91-45904
516--dc20 CIP

Printed on acid-free paper.

ISBN 0-8176-3620-X
ISBN 3-7643-3620-X

Typeset by Ark Publications, Inc., Newton Centre, MA
Printed and bound by Quinn-Woodbine, Woodbine, NJ
Printed in the U.S.A.

9 8 7 6 5 4 3 2 1

Dedicated to the Memory of Cyril S. Smith, Supreme Philomorph.

Preface

Concepts and Images is the result of twenty years of teaching at Harvard's Department of Visual and Environmental Studies in the Carpenter Center for the Visual Arts, a department devoted to turning out students articulate in images much as a language department teaches reading and expressing oneself in words. It is a response to our students' requests for a "handout" and to our colleagues' inquiries about the courses[1]: Visual and Environmental Studies 175 (Introduction to Design Science), VES 176 (Synergetics, the Structure of Ordered Space), Studio Arts 125a (Design Science Workshop, Two-Dimensional), Studio Arts 125b (Design Science Workshop, Three-Dimensional),[2] as well as my freshman seminars on Structure in Science and Art. *Concepts and Images* is designed to be used in conjunction with the first and third of these courses, whereas *Space Structures, their Harmony and Counterpoint* in the Design Science Collection covers topics from the other two courses, being primarily three-dimensional.

In these courses I work to overcome visual illiteracy and mathematics anxiety, two serious and related problems. Visual illiteracy affects our man-made environment and its relation to our natural ecology. Mathematics anxiety deprives those afflicted of access to the grammar needed to express oneself spatially. The visual clutter of our built environment and the sterility of ultra-functionalist design reflect the intellectual poverty of our environment. Only a proper understanding of the constraints imposed by the properties of our space and of the rich repertoire permitted within these contraints allow the achievement of a balanced, disciplined freedom.

Within our culture modern natural sciences and mathematics have become less accessible because of the specialized skills required to become conversant with them. Nevertheless, in piling discovery upon discovery, we do not usually follow the most direct path to a concept, and if we were to retrace our steps we might well recognize that not all the byways by which such a concept was developed historically are, in point of fact, prerequired for understanding the concept. Although such retracing requires time and effort, it opens up channels of communication which may enrich our culture.

The four courses and the seminars mentioned above could be considered to constitute a *laboratory for visual mathematics*: Experiments with particular patterns lead inductively to more generally applicable fundamental principles. In *Concepts and Images* mathematics is introduced as needed: The reader should merely be willing to abandon some previously acquired notions not based on actual observation and to arrive at some abstract ideas inductively from the experiments suggested in the text. Artistic skills are not required beyond a willingness to express oneself graphically as articulately and elegantly as one's skills permit. For this reason the book is designed as a workbook; illustrations may easily be traced, and the text is full of questions *PRINTED IN UPPER CASE ITALIC TYPE*, for I have found that it is more difficult to ask the right questions than it is to find the correct answers. These questions will provide the motivation for further exploration and the acquisition of further skills. The illustrations were produced with the aid of neither computer graphics nor of skilled draughtsmen, because it was felt that they should match the style which can be attained by the average reader with the use of straight edges, triangles, compasses and the French curve.

The balance between order and disorder, referred to above as *disciplined freedom*, already impressed me as a boy when I watched my grandfather assemble his collection of antique Delft tiles, at present in the Rijksmuseum in Amsterdam. My earlier *Color and Symmetry*[3] was dedicated to his memory. The subtle spatial rhythm resulting from nearly, but not perfectly, identical hand-made tiles is quite different from that of an assembly of perfectly cloned machine-made tiles. Just as a string in equilibrium is silent, but when vibrating around equilibrium may emit a most ravishing sound, so a too perfect symmetry is silent, and vibrant only when gently perturbed. So-called mathematical perfection is an abstraction not *achieved* but nevertheless *perceived* in a concrete representation. It is hoped that the illustrations in this book will clarify the mathematical abstraction without intimidating by having too mathematical a perfection.

There is much to complain about in *Concepts and Images*. Mathematicians may find it trivially elementary or insufficiently rigorous; artists may be intimidated by its abstractions. Yet it is hoped that there will be mathematicians who will be stimulated to adopt the visual mathematics approach in their own teaching, as I have already tasted the satisfaction of viewing fine art from artists who had their vision broadened by new concepts acquired through design science.

What is included in and excluded from *Concepts and Images* is principally determined by the response from my students in my courses and freshman seminars, and it is to these students that the book is dedicated with sincere appreciation. My assistants, Holly Compton Alderman, William L. Hall, Jack C. Gray, William Varney, P. Frances Randolph and Caryn Johnson have each contributed in their individual ways to the shaping of the courses and have all left their mark on this book. I am indebted to Jack Gray in particular for keeping our bibliography up-to-date and for recording his responses to student work.

The Rockefeller Foundation, in granting me a residency at their incomparable facility in the Villa Serbelloni on Lake Como, enabled me to draw the illustrations and to edit the manuscript. The considerate care which the late and lamented Roberto Celli and his lovely wife Gianna lavished on us will never be forgotten. The feedback from my fellow residents, in particular Joan Krieger and Petah Coyne, who showed me the riches of computer-aided and hand-drawn graphics, was invaluable.

Cambridge, Massachusetts

NOTES

[1] These courses were crosslisted in the Graduate School of Design and in the Graduate School of Education under various numbers, among them Arc 1-9a and 1-9b.

[2] The latter two offered in the Faculty of Arts and Sciences' Harvard Extension Program.

[3] Loeb, A. L.: *Color and Symmetry.* Wiley, New York (1971); Krieger, New York (1978).

I

Introduction

What is Design Science?

Just as the grammar of music consists of harmony, counterpoint and form (sonata, rondo, etc.) which describe the structure of a composition and poetry has its rondeau, ballad, virelai and sonnet, so spatial structures, whether crystalline, architectural or choreographic, have their grammar, which consists of such parameters as symmetry, proportion, connectivity, valency, stability. Space is not a passive vacuum; it has properties which constrain as well as enhance the structures which inhabit it. Design science comprises this grammar in the broadest sense, dealing with those parameters which are common to all spatial structures. That this grammar did not become obvious earlier is probably due to the fact that crystallographers, architects, mathematicians, visual artists and choreographers have worked on such different scales and in such varied idioms that they found it hard to communicate.[1]

Ours is an age of images. Signs and images are becoming more potent as the prevailing means of public communication.[2] Generations educated under the influence of television are not only receiving much of their early impressions in the form of images rather than words; they also will tend to form their concepts in terms of images rather than words. It may therefore be erroneous to believe that recent generations of students are less "literate" than their predecessors if this literacy is only measured as verbal literacy. Not only are we communicating by means other than linear strings of words, we are also developing different ways of thinking, of articulating our ideas and of solving problems.

We have reason to believe that intuition is a form of non-verbalized knowledge.[3] R. Buckminster Fuller called one of his books as well as one of his boats "Intuition." Fuller is known for his images, primarily his dome structures and his tensegrities; his difficulty in communicating verbally is characterized by the special nomenclature he needed to create to express himself in words.

Judith Wechsler,[4] in a series of articles by well-known natural scientists, has demonstrated that the path to scientific discovery does not follow the same orderly course implied by the eventual publication of the results of the discovery. The path of discovery does not begin with an answer; rather, discovery depends on a question. All learning starts with questions, not answers. It is usually more difficult to formulate the right question than to find the answer. Only by understanding a question will a student understand why certain skills need to be learned and hence feel motivated to exert a certain effort to acquire these skills. It is not easy to use this method in a book, which unfortunately needs by and large to be a linear string of words. Nevertheless, the reader will find the text interspersed with *QUESTIONS PRINTED IN ITALIC UPPERCASE LETTERS*. This means that the reader should attempt to find an answer before reading further; thinking about the question posed will lead into succeeding paragraphs. This book should be used interactively: It will prove helpful to have straight edge, pencil and compass on hand.

In discussing the lack of linear progression in most scientific discoveries, I by no means belittle the importance of a rational and systematic approach to problems; on the contrary, this book is full of examples where mathematical reasoning provides answers when intuition would leave us perplexed. The two approaches complement each other and should work synergetically. Since in most of us either the analytical or the intuitive skill is dominant, we need to train the other to work with the dominant one.

To illustrate the distinction between intuitive and analytical approaches, consider these two figures.

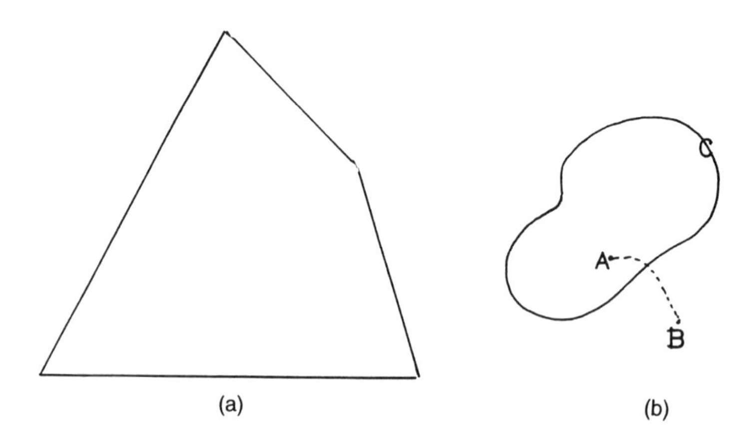

Figure 1-1 (a) and (b) ***Tile and Circuit***

The configuration on the left represents a tile whose identical copies are to be used to cover a very large flat area without overlapping or leaving any space in between. Such a tiling without either gaps or overlaps is called a tessellation.

Q: *WOULD A QUADRILATERAL TILE AS SHOWN IN FIGURE 1-1 (a) BE SUITABLE FOR COVERING A PLANE WITHOUT OVERLAP OR INTERSTICES?*

Intuitively it is not at all obvious that this can be done. While you ponder this problem, look at Figure 1-1 (b), which represents a closed circuit, C, on a surface. Two points, A and B, are located on the same surface, A inside circuit C, B outside.

Q: *WILL ANY LINE ON THE SAME SURFACE WHICH JOINS POINTS A AND B NECESSARILY CROSS CIRCUIT C?*

Intuitively one would not doubt that this must be so, but for many centuries the proof was a vexing one in graph theory.

We could accept as a basic postulate the fact that on a surface a line joining a point inside a closed circuit to a point outside that circuit necessarily crosses the circuit. Such an assumption would not cause us much concern unless we would encounter an instance which would violate this postulate. Nevertheless, the mathematician, wishing to minimize the number of basic postulates needed for a self-consistent system, will not readily accept the necessity for adding a new one, no matter how obvious the truth of the new postulate may appear intuitively.

Let us now return to Figure 1-1 (a). Trace the outline of the tile on a piece of paper. Mark the four midpoints of the respective edges of the quadrilateral. Rotate the quadrilateral 180° around each of the centers of edges just marked, tracing each new quadrilateral so marked (see Figure 1-2). Repeating the process, marking the midpoints on the edges of each newly generated quadrilateral, we create the tessellation in Figure 1-3.

The fact that this, and hence *any* irregularly shaped quadrilateral which has straight edges, can fill the plane together with identical copies of itself without any leftover spaces or overlaps is almost counter-intuitive. Thus we have encountered two contrasting examples: one whose conclusion was intuitively evident but mathematically very difficult to prove, and another which could easily be proven mathematically but which was intuitively not at all obvious.

Does this mean that the objectives of the mathematician and of the designer are at variance with each other? On the contrary: The mathematician can help the designer immeasurably by expanding the latter's repertoire beyond what is intuitively evident. In turn, the scientist is inspired by matters of personal

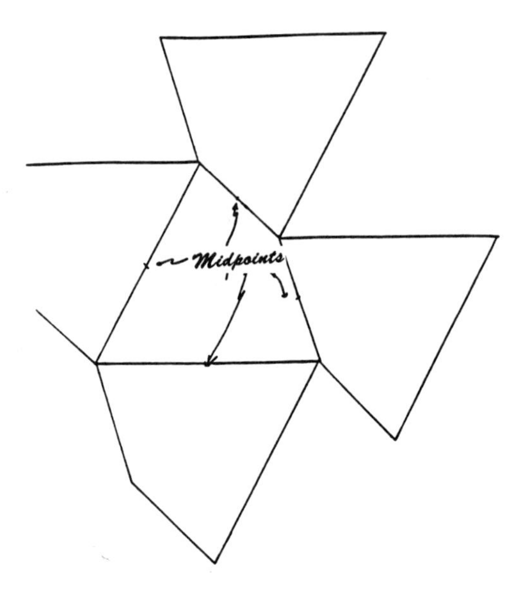

Figure 1-2 ***Tiling the plane with an arbitrary quadrilateral***

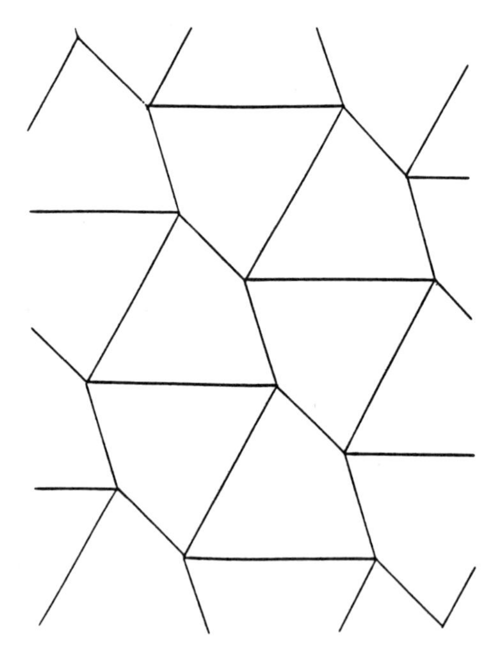

Figure 1-3 ***Quadrilateral tessellation***

taste and beauty and uses intuition in the process of discovery (Judith Weschler, *op. cit.*).

The objective of this book is to combine analysis and intuition to deal with spatial complexities and to explore the process of discovery through experiments and the development of generalized concepts by finding the principles underlying the results of these experiments.

The first section of the book deals with the symmetry of discrete systems, essentially the continuation of the tiling problem just discussed. The last part of this book is concerned with dynamic symmetry and spirals. These spirals are best appreciated with the aid of calculus; in point of fact, spirals are a fine visual introduction to calculus. Because calculus is a stumbling block for many who are in the field of design science, I have not assumed any knowledge of it, but instead give, in the middle section of the book, a kind of calculus tutorial using logarithmic spirals as an example. Those fluent in calculus may skip this section without problem, although they may find it useful in teaching calculus themselves.

NOTES

[1] Chorbachi, Wasma' a: *In the Tower of Babylon: Beyond Symmetry in Islamic Design* in *Symmetry 2, Unifying Human Understanding,* Istvan Hargittai, ed. Pergamon, New York (1989).

[2] Young, Canon Jonathan F., in "The Lantern, A Monthly Newsletter of the Old North Church." Boston, Massachusetts, January 1990.

[3] Haughton, E. C., and A. L. Loeb: *Symmetry, the Case History of a Program,* J. Res. in Science Teaching, **2**, 132–145 (1964).

Loeb, A. L., and E. Haughton: *The Programmed Use of Physical Models,* J. Progr. Instr., **III**, 9–18 (1965).

[4] Wechsler, Judith, ed.: *On Aesthetics in Science.* MIT Press, Cambridge, MA. (1978), reprint Birkhäuser Design Science Collection, A. L. Loeb, series ed., Birkhäuser, Boston (1988).

II

Areas and Angles

Q: *WHAT IS THE AREA OF A SQUARE WHOSE EDGE-LENGTH EQUALS ONE CENTIMETER?*

Someone may have told you that it equals one square centimeter. Did you ever see a proof of this statement? If not, do you suppose that it was a definition?

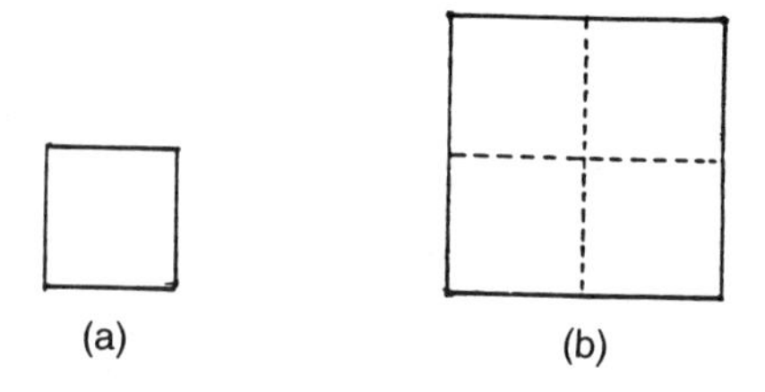

Figure 2-1 (a) and (b) ***Comparison of two squares***

In Figure 2-1 we show two squares, one of which has edges twice as long as the other does. The area of Figure 2-1 (b) is clearly four times as large as that of Figure 2-1 (a). In Figure 2-2 there are two triangles, geometrically similar to each other.

Q: *WHAT IS THE RATIO OF THE LENGTHS OF THE EDGES OF THE TWO TRIANGLES? WHAT IS THE RATIO OF THEIR AREAS?*

The edge length of one triangle is three times that of the other, and we note that the triangle in Figure 2-2 (b) has an area nine times as large as that of Figure 2-2 (a).

The two squares in Figure 2-1 are geometrically similar, as are the triangles in Figure 2-2. For each point in the smaller figure of each pair there is a corresponding point in the larger: The upper left vertex of the smaller square corresponds to the upper left vertex of the larger square; the intersection of the diagonals of one square corresponds to that of the other square; the center of the base of one triangle corresponds to the center of the base of the other triangle. A line segment joining a pair of points in one figure corresponds to the line segment joining the corresponding points in another figure geometrically similar to the first.

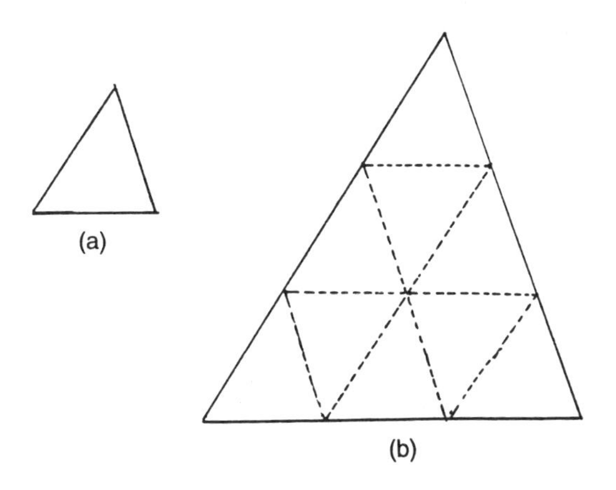

Figure 2-2 (a) and (b) ***Comparison between triangles***

Figures 2-1 and 2-2 demonstrate

Theorem 2-1: The ratio of the areas of two geometrically similar configurations equals the *square* of the ratio of the lengths of corresponding line segments.

For instance, the areas of the squares in Figure 2-1 are in the same ratio as the square of the lengths of their diagonals as well as of their edges. And the ratio of the areas of the two triangles in Figure 2-2 equals the square of the ratio of the lengths of their bases and also of the distances between their upper vertices and their bases. Nothing can be concluded about the *absolute* values of the areas, only about their ratios: If l_1 and l_2 are the lengths of corresponding line segments of two geometrically similar configurations, and A_1 and A_2 are their areas, then:

$$\frac{A_1}{A_2} = (\frac{l_1}{l_2})^2, \quad \text{or} \tag{2-1}$$

$$A = Cl^2, \tag{2-2}$$

where C is a proportionality constant which depends solely on the *shape*, and not on the *scale*, of the configuration.

The area A of a square of edgelength l thus is given by $A = Cl^2$, not necessarily $A = l^2$. Whoever asserts the latter arbitrarily sets $C = 1$ for a square, but in doing so arbitrarily assigns fundamental significance to the square. With equal justification, and in some cases more usefully, one could set $A = l^2$ for an equilateral triangle having edgelength equal to l.

Since all squares are mutually similar, all have the same shape factor C. Lengths, areas and volumes are as different from each other as are forces, distances and electrical charges. Coulomb's Law expresses the force, F, between two charges, q_1 and q_2 that are separated by a distance r:

$$F = \frac{Cq_1 \cdot q_2}{r^2},$$

where C is a proportionality constant, which depends on the units chosen for force, distance and electrical charge as well as on the medium within which the force operates. There are several systems of units for electrical charges, just as there are different units for distance (feet, meters, lightyears, etc.), which determine the value and units of the constant C. A particular set of electrical charge units is chosen such that the value of the proportionality constant C is unity *in vacuo*; this is an arbitrary choice, just as is the choice of the proportionality constant $C = 1$ for a square. In the case of Coulomb's Law, just as in the law relating area to length, we can only measure how one variable changes when another one changes: We can compare areas corresponding to different lengths, or forces between charges of different magnitude at different distances. We cannot, however, make *absolute* measurements of magnitude.

Equations (2-1) and (2-2) are experimentally verifiable expressions. The relationship $A = l^2$ is not: It results from an arbitrary assumption that $C = 1$ for the square, and it has left generations of architects, designers and others with the mistaken impression that the square is somehow unique and fundamental. The square does not occur frequently in nature, nor is the right angle encountered frequently in the art of people who are much in touch with nature. We shall not depend on the assumption $C = 1$ for the square; we instead base our relationships on equations (2-1) and (2-2), returning to these when we consider volumes of three-dimensional bodies.[1]

To demonstrate the adequacy and power of equations (2-1) and (2-2), which of course are two different expressions of the same relationship, we shall derive from them two other important relations, the first being the area of a rectangle having width w and height h (Figure 2-3). Construct a square on the long side of the rectangle and a second one on the short side (Figure 2-4). This construction implies another rectangle w by h, perpendicular to and sharing a vertex with the original one. The two squares and the two rectangles add up to a larger square of the edgelength $(w + h)$, whose area is $C(w + h)^2$. Since the areas of the two smaller squares are Cw^2 and Ch^2, the areas of the two mutually identical rectangles add up to $C[(w + h)^2 - (w^2 + h^2)]$, with the result that the area of each rectangle equals $\frac{1}{2}C[w^2 + 2wh + h^2 - w^2 - h] = Cwh$. While this result is hardly surprising, it is reassuring that the expression for the area of a rectangle may be derived from that for the area of a square: If we should arbitrarily call the area of a square having edgelength l, $A = l^2$, then the area of a rectangle having width w and height h will be $A = wh$.

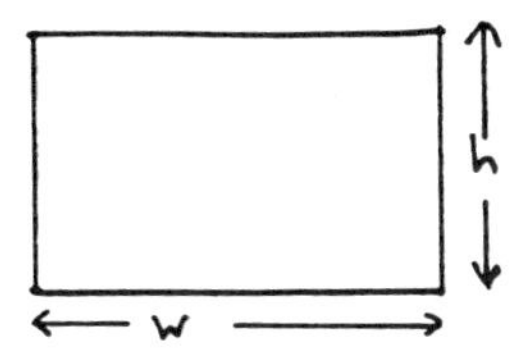

Figure 2-3 ***A rectangle***

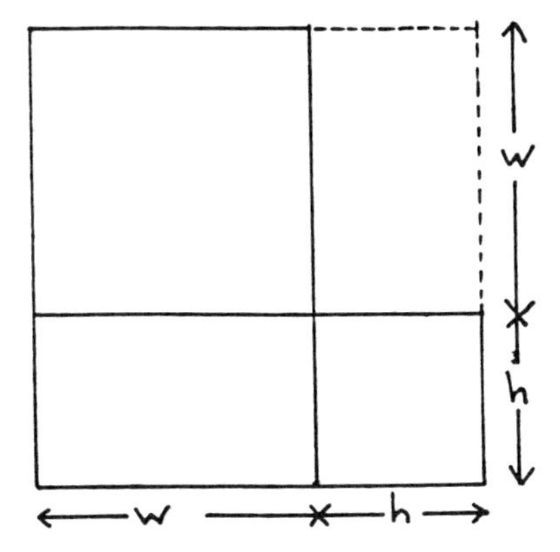

Figure 2-4 ***Derivation of the formula for the width of a rectangle from equations (2-1) and (2-2)***

From rectangle to parallelogram is a simple step: Move a small triangle from one side of the rectangle to the opposite one (Figure 2-5). The equation of the area of a triangle is obtained from that of a parallelogram by diagonal bisection (Figure 2-6). Accordingly, the area of a triangle is $\frac{1}{2}Cwh$; the factor one-half simply represents the bisection of the parallelogram.

Q: *WHAT WOULD BE THE AREA OF A SQUARE IF THE AREA OF AN EQUILATERAL TRIANGLE HAVING UNIT EDGELENGTH WERE TO BE SET AT UNITY?*

There are occasions when it is useful to choose as a unit area that of an equilateral triangle of unit edgelength; its height would be $h = \frac{1}{2}\sqrt{3}$, hence its area would be $\frac{1}{4}C\sqrt{3}$. If this is to equal unity, then C must be chosen to be $4/\sqrt{3}$; in such a system the area of a square would be $(4/\sqrt{3})l^2$.

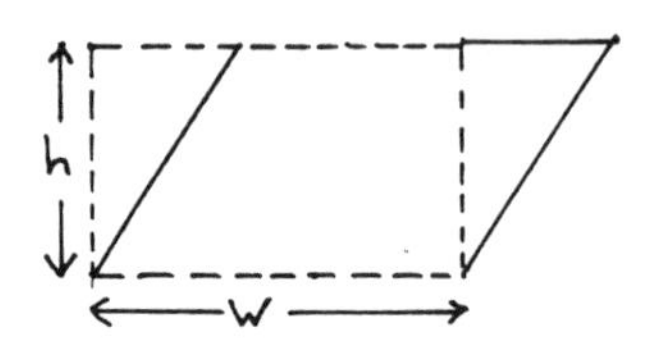

Figure 2-5 ***Area of a parallelogram***

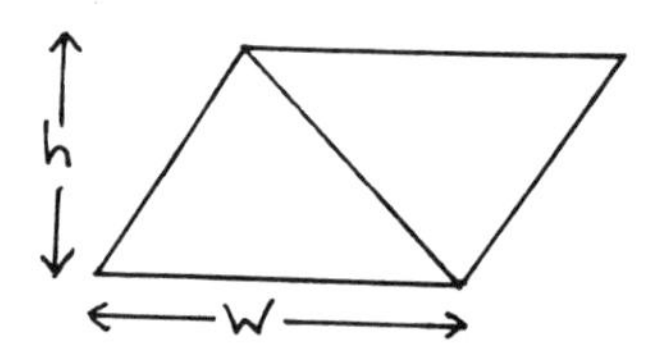

Figure 2-6 ***Area of a triangle***

A second application of Equation (2-2) is as follows. Consider in Figure 2-7 a right triangle, ABC, C being the right-angled vertex. Drop the perpendicular from C onto the hypotenuse AB, calling their intersection E. Applying equation (2-1) to right triangles, we can say:

$$\text{Area (ABC)} = K \cdot AB^2$$

$$\text{Area (ACE)} = K \cdot AC^2$$

$$\text{Area (BCE)} = K \cdot BC^2$$

where K is the shape constant for all three triangles, which are mutually similar. Since

$$\text{Area} \quad (ABC) = \text{Area} \quad (ACE) + \text{Area} \quad (BCE), \quad \text{then}$$

$$AB^2 = AC^2 + BC^2 ,$$

so that we have derived Pythagoras's theorem directly and simply from the fundamental equation (2-2).

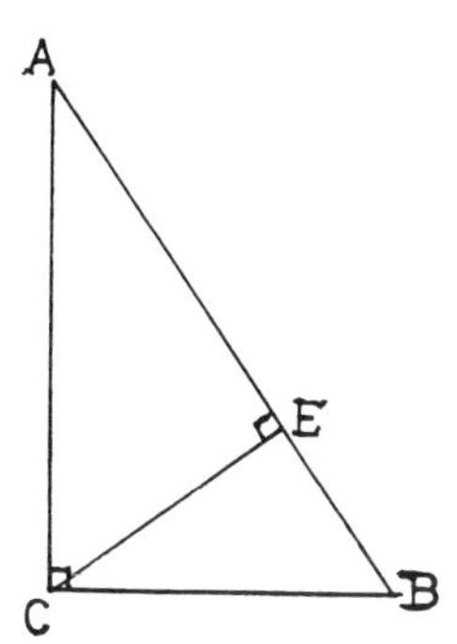

Figure 2-7 ***A right triangle***

Another parameter with which we shall be concerned is the *angle*, a measure of direction. Angular measurement is different from distance, area or volume in that we have here the opportunity for an unequivocal universally acceptable unit: The concept of a single complete revolution does not depend on locally idiomatic ideas or languages. With the use of a circular disc it is not difficult to introduce the notion of equal angles and of dividing an angle into parts, even though the actual construction, say the trisection of an angle, may present problems. In geometrically similar figures corresponding angles are all the same, so that scaling does not affect angles as it does distances.

In principle, angles could be expressed as fractions of a complete revolution: A right angle would have the value 1/4, an equilateral triangle would have angles equal to 1/6. In practice, to avoid fractional values, a very small angle, equal to 1/360 of a revolution, was used as a unit and assigned the value of 1 *degree*. The number 360 is useful for this purpose, because it is divisible by 2, 3, 4, 5, 6, 8, 9, 10, 12, 15, 18, 20, 24, 30, 36, 40, 45, 60, 72, 90, 120, 180, so that a multitude of angles is expressible as an integer multiple of degrees. Yet it is a practical rather than a fundamental unit, and equations expressed in terms of degrees are not always convenient or elegant. In Figure 2-8 are two circular arcs, AB and CD, having a common center, O. If the line segment AO were half the length of CO, then the length of arc AB would be half that of arc CD. Generally, the arc length L is proportional to the radius R:

$$L = \theta R, \tag{2-3}$$

where θ is a proportionality constant determined by the angle subtended by the two arcs.

Since this proportionality constant is independent of scale and only depends on the difference in directions between lines OAC and OBD, it is in point of fact a measure of the angle between these two lines and may therefore be called the angle between these lines. A *unit angle*, then, is the angle subtended by a

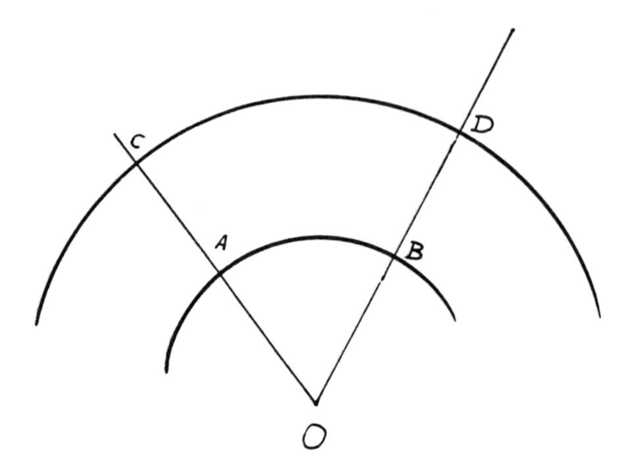

Figure 2-8 ***Circular arcs***

circular arc whose length equals that of its radius. Such an angle is called a *radian.* Equation (2-3) may be used to relate arclength to angle subtended and to radius, as long as the angle is expressed in radian units.

Q: *DRAW AN EQUILATERAL TRIANGLE OAB AND A CIRCULAR ARC AB WHOSE CENTER IS AT O. OF THE FOLLOWING THREE ELEMENTS, WHICH IS THE LONGEST: THE LINE SEGMENT AB, THE LINE SEGMENT OA, OR THE ARC AB? ARRANGE THE FOLLOWING ANGLES IN INCREASING ORDER OF MAGNITUDE: 60°, π RADIANS, 1 RADIAN, 45°, 180°.*

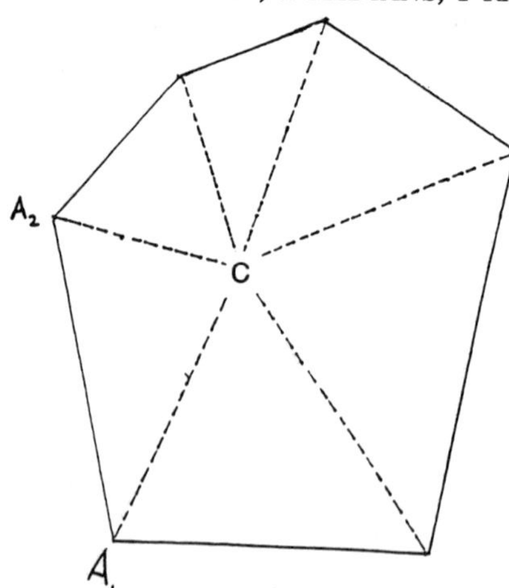

Figure 2-9 ***A polygon***

Since the arc subtended by a complete circle (a complete revolution or 360°), has length 2π times the radius, 360° $= 2\pi$ radians. Since the number π is a little more than 3, one radian is a little less than 60°.

Figure 2-9 shows a polygon, a closed circuit composed of (in this instance) straight line segments, called *edges*. In general, a polygon has n edges, two adjacent ones joined at a vertex. Accordingly, the polygon has n vertices as well. To find the sum of the (internal) angles of an n-edged polygon, choose a point C inside the polygon, and join it to the vertices labeled successively A_1, $A_2, \ldots, A_n$. The sum of the angles of a triangle being 180°, i. e., π radians, the sum of the angles of triangle A_1 A_2 C equals π radians, as is the sum of the angles of triangle A_2A_3C, etc. The polygon has n such triangles, whose angles will add up to $n\pi$ radians. These angles include 2π radians around the point C; therefore the internal angles of the n-edged polygon add up to $(n-2)\pi$ radians. For a *regular* polygon, that is, one whose edges are all equal in length and whose angles are all identical, each integral angle equals $(1 - 2/n)\pi$ radians. For example, an equilateral triangle has $n = 3$, hence its internal angles are $\pi/3$ radians; for a square, $n = 4$, so that the internal angle is $\pi/2$ radians; for a regular pentagon, $n = 5$, and the internal angle equals $3\pi/5$ radians, or 108°; and for a regular 100-gon, the internal angle would be $49\pi/50$ radians, or 176.4°.

NOTES

[1] Loeb, A. L.: *Remarks on Some Elementary Volume Relations between Familiar Solids,* The Math. Teacher, **LVIII**, 417–419 (May, 1965).

Loeb, A. L.: Preface and Contributions to R. Buckminster Fuller's *Synergetics*, 832–836. MacMillan, New York (1975).

III

Tessellation and Symmetry

In Chapter I we covered the plane with a set of identical irregular quadrilaterals. In general, a surface which is entirely covered by polygons without overlap or spaces in between is said to be *tessellated*, a term derived from the Greek word *tessella*, or tessera, a tile. Principally, we shall concern ourselves here with the problem of covering a plane with mutually identical tiles; in Chapter XX we shall deal with a particular *pair* of tiles.

Although mosaic art is about as ancient as our civilization, it has gained new relevance in our age of mass production. Indeed, there are a few technical obstacles to creating a free-form tile, but conceptually it is more difficult to figure out which shapes will fit together to cover the plane when the tile is anything but a rectangle. Mass production is not the only reason for tiling surfaces with the use of identical modules: The rhythms created by the repetition of identical patterns have a fascination of their own. Islamic art abounds with such visual rhythms [1,2,3]; the twentieth-century Dutch artist M. C. Escher acknowledged the influence of Alhambra on his own periodic designs.

Figure 3-1 shows a tessellation using mutually identical equilateral triangles.

Q: *COULD YOU TESSELLATE THE PLANE WITH ANY TRIANGLE WHICH IS NOT REGULAR?*

(Remember: A *regular* triangle has all edgelengths the same and all angles identical.)

It is not necessary to limit oneself to equilateral triangles: Any triangle having straight edges will tessellate the plane (Figure 3-2). Triangles do not necessarily have straight edges; those in Figure 3-3 do not.

Q:*WHICH CURVED-EDGED TRIANGLES WOULD, AND WHICH WOULD NOT, TESSELLATE?*

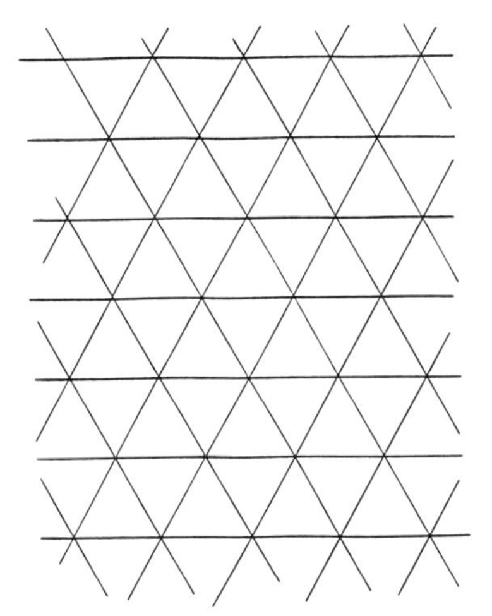

Figure 3-1 ***Triangular tessellation: Regular triangles***

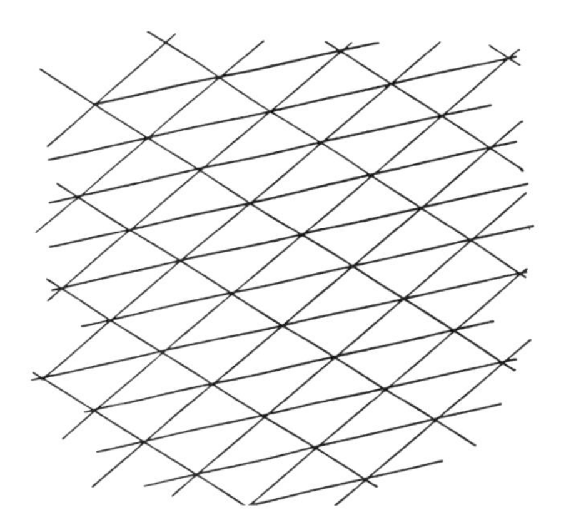

Figure 3-2 ***Triangular tessellation: irregular triangles***

To find out, let us first realize that any parallelogram tessellates the plane (Figure 3-4). In this tessellation, each parallelogram is in exactly the same orientation as all its neighbors; the same cannot be said for the triangular tessellations of the first two figures. Note that Figures 3-1, 3-2 and 3-4 have the following feature in common. Let us assume that they extend indefinitely in all directions, and let us trace a finite portion of the pattern on a transparent piece of paper. Then move the transparent paper *parallel to itself.* When the paper is moved through certain distances in certain directions, the pattern traced on it will exactly overlay that seen underneath. A pattern that has this property is said to have *translational* symmetry.

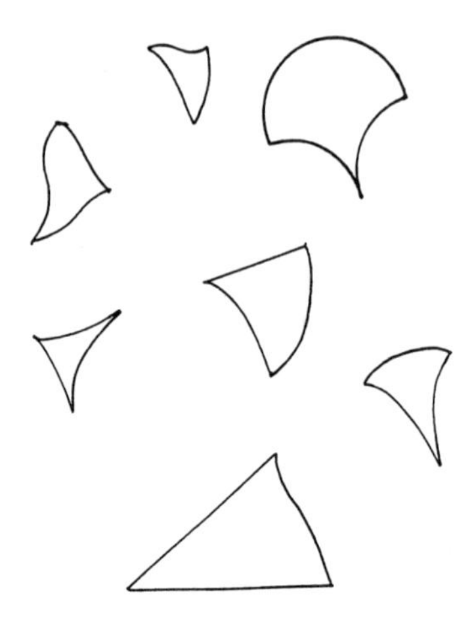

Figure 3-3 ***Triangles having curved edges***

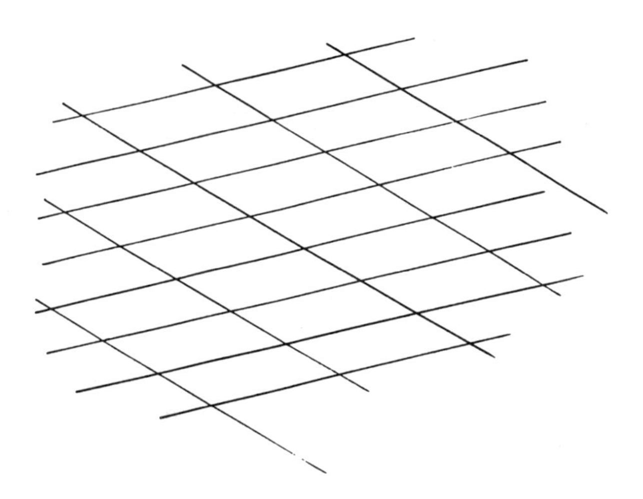

Figure 3-4 ***Parallelogram tesselation***

Translational symmetry thus expresses the periodic repetition of a motif in a pattern, this motif being always in the same orientation (Figure 3-5).

Q: *WHICH TRIANGLES IN FIGURES 3-1 AND 3-2 ARE RELATED BY TRANSLATIONAL SYMMETRY?*

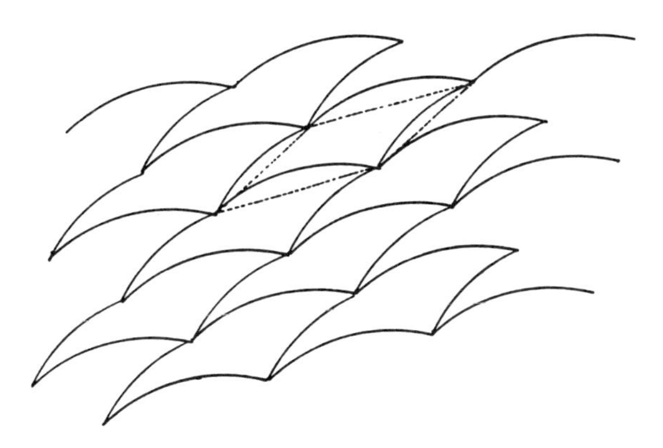

Figure 3-5 ***Translational symmetry: Curved-edged parallelograms***

The triangles in Figures 3-1 and 3-2 are in two different orientations, half of them having their base above the third vertex, the others being inverted compared to the first half. Although Figures 3-1 and 3-2 do have translational symmetry, the parallel motion of the tracing paper will not bring upright triangles into coincidence with inverted ones. The latter effect can be achieved by placing a pin exactly in the middle of any side of any of the triangles, and rotating the pattern 180° around the pin. Any pattern that can thus be brought into coincidence with itself by a rotation around a point is said to have *rotational symmetry*. The point is called a *center of rotational symmetry*, or, abbreviated, a *rotocenter*.

Figure 3-5 has translational, but not rotational, symmetry. The motif is a parallelogram having curved sides, that is to say, its sides are pairwise parallel to each other, but curved. This pattern could have been derived from a tessellation with straight-edged parallelograms by scooping out a portion from one side of the parallelograms and adding this portion to the opposite sides. We may conclude that a quadrangle (a polygon having four sides) will certainly tessellate the plane if opposite sides are mutually congruent and parallel.

Q: *ARE THERE QUADRILATERALS WHICH TESSELLATE THE PLANE BUT DO NOT HAVE MUTUALLY CONGRUENT PARALLEL OPPOSITE EDGES*?

This is a sufficient, but not a necessary condition, for, as we may have already discovered, there are many quadrangles which do not meet this criterion, and nevertheless tessellate the plane (cf. Figure 1-2).

The notion that a polygon does not necessarily have straight edges may come as a surprise. It forces us to define just what we mean by the terms *vertex* and *edge*. A *vertex* is a point; it has no dimension. An edge has a single dimension: Although it is not necessarily a straight line, at any one point on it one can travel along it (either forward or backward) in only one direction. An edge contains only two vertices, one at each end: An edge joins two vertices, but does not contain any other vertices. There is no reason why we should not place a vertex at a location between the two terminal vertices of an edge, but in doing so we will have divided the edge into two edges, each having a vertex at either end (Figure 3-6). An n-sided polygon has n edges and n vertices, which alternate when one travels around the circumference of the polygon.

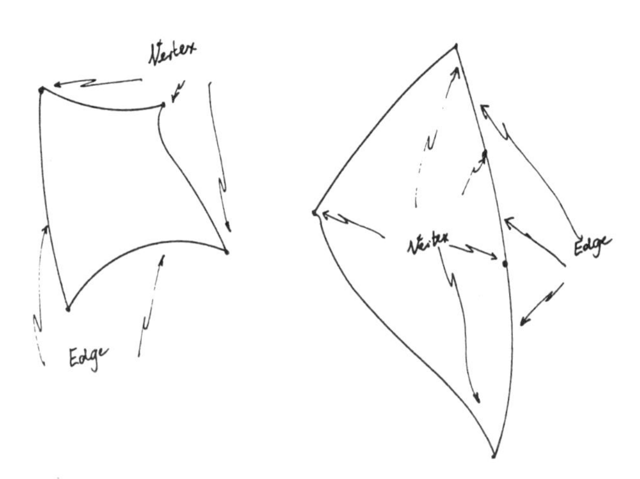

Figure 3-6 ***Edges and vertices***

The polygons in a tessellation are called *faces*. A *face* is enclosed by circumferential vertices and edges but does not contain any internal ones. Whereas a vertex has no dimensions and an edge is one-dimensional, a face is two-dimensional, even though it is not necessarily flat: One may tessellate non-planar surfaces, for instance the surface of a sphere.[4]

Returning to the question of which of the triangles having curved edges will tessellate the plane, we note that the straight-edged tessellation of Figure 3-2 contains many centers of rotational symmetry.

Q: *HOW MANY DISTINCT CENTERS OF ROTATIONAL SYMMETRY, THAT IS TO SAY, CENTERS WHOSE CONTEXTS IN THE PATTERN ARE DIFFERENT, ARE THERE?*

All the vertices constitute such centers, as do the centers of all edges. All these centers have two-fold rotational symmetry: A complete 360° rotation of the entire pattern around such a center will reveal the same pattern exactly *twice* in the same orientation. Thus there are four different kinds of centers of two-fold rotational symmetry. If we distort the straight edges while maintaining their two-fold rotational symmetry, we generate a curved-edge triangular tessellation (Figure 3-7).

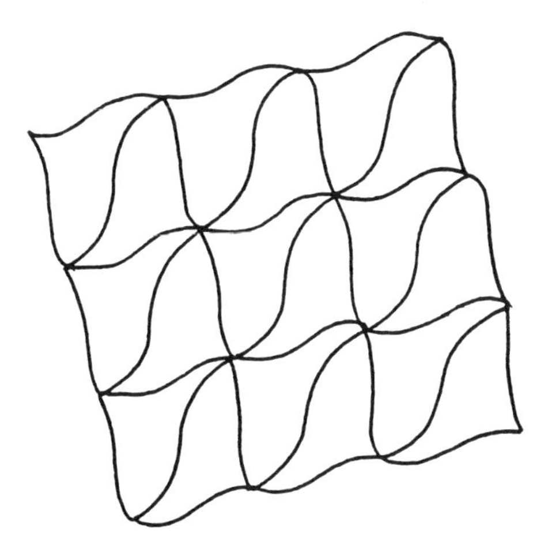

Figure 3-7 ***Curved-edged triangular tessellation having two-fold rotational symmetry***

It is easy to see that the curved-edged tessellation by parallelograms of Figure 3-5 possesses no rotational symmetry, in contrast to the straight-edged tessellation of Figure 3-4. In the latter, each parallelogram may be bisected by either one of two diagonals. One such division generates Figure 3-2, the other is shown in Figure 3-8. In either case a straight-edged triangular tessellation is generated, having two-fold rotational symmetry in exactly the same locations as did the original parallelogram tessellation, namely at all vertices, at the centers of all edges, and in the centers of all faces of the parallelograms. By contrast, the curved-edged parallelograms in Figure 3-5 cannot be subdivided into two mutually congruent halves. On the other hand, pairs of adjacent curved-edged triangles in Figure 3-7 may be joined together to form curved-edged parallelograms. Accordingly, these will tessellate, resulting in a tessellation which, in contrast to that of Figure 3-5, does possess two-fold rotational symmetry. The reason is that a tessellation by parallelograms formed from Figure 3-7 has centers of two-fold rotational symmetry at the midpoints of each of its edges.

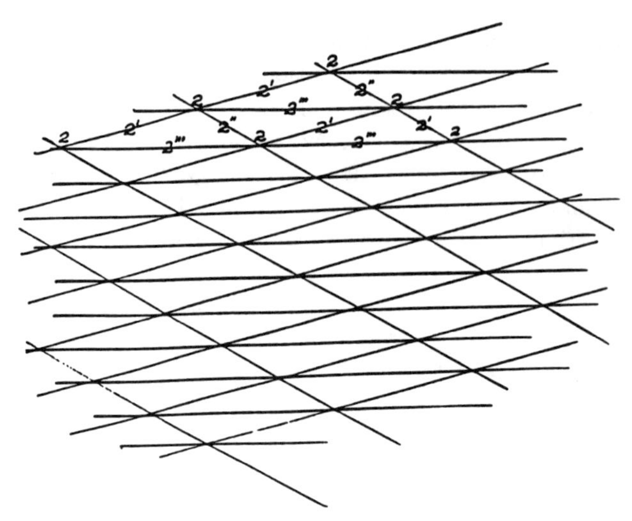

Figure 3-8 ***Subdivision of a straight-edged parallelogram tessellation into a triangular tessellation***

Q: *WOULD SUCH A TESSELLATION HAVE ALL ITS CENTERS OF ROTATIONAL SYMMETRY ON THE EDGES OF ITS TILES? IF THERE ARE ADDITIONAL ONES, WHERE WOULD THEY BE LOCATED?*

In each of these tessellations having two-fold rotational symmetry the rotocenters lie equally spaced along parallel rows; here we have used primes to distinguish rotocenters whose contexts in the pattern are different. This configuration is a direct result of their two-fold symmetry: If two two-fold rotocenters 2 and $2'$ coexist (Figure 3-9 (a)), then the symmetry of 2 requires a second $2'$ on the other side of 2 (Figure 3-9 (b)), whereas the symmetry of $2'$ requires a second 2 on the other side of $2'$ (Figure 3-9 (c)).

Q: *CONTINUING WITH THE SAME REASONING, HOW MANY ROTOCENTERS WOULD YOU GENERATE?*

In turn, these new two-fold rotocenters imply additional rotocenters on the same straight line, with the result that an infinitely large number of rotocenters is implied (Figure 3-9 (d)). This linear array of two-fold rotocenters would correspond to a frieze such as illustrated in Figure 3-10, constructed out of curved-edged triangles. To generate a tessellation of the entire plane, however, it will be necessary to engage a two-fold rotocenter which does not lie on the same line as did 2 and $2'$.

Such a rotocenter, called $2''$, will imply a second row of equally spaced two-fold centers equivalent to the first one, located at the opposite side of

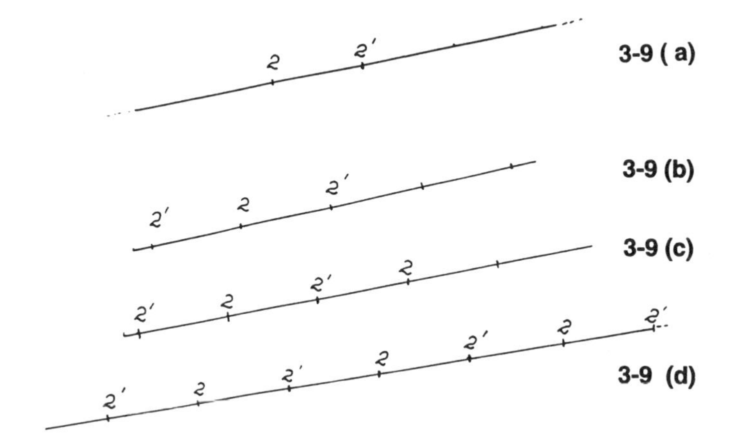

Figure 3-9 ***Generation of a row of two-fold rotocenters***

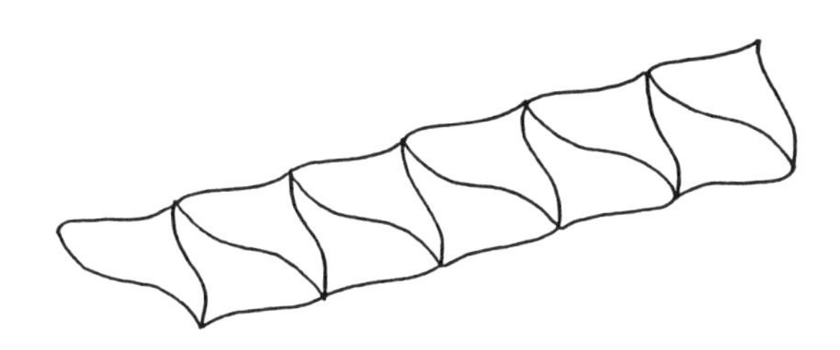

Figure 3-10 ***Frieze created out of curved-edged triangles***

$2''$ (Figure 3-11). If we look at the tessellations by curved-edged triangles in Figure 3-7, we recall that there are *four* sets of distinct rotocenters, distinct meaning that they have different contexts in the pattern. It is evident that there are $2'''$ rotocenters as well, as an inevitable result of the other three sets.

Q: *INDICATE THE LOCATION OF THESE ROTOCENTERS IN FIGURE 3-11. CHECK YOUR RESULT WITH FIGURE 3-12.*

We have come a considerable way from the straight-edged rectangle in our search for a free-form tessella. Above we cautioned that in addition to parallelograms there may be other quadrilaterals which tessellate as well. Using the general straight-edged quadrilateral of Figure 3-13 as a module, we can generate the tessellation of Figure 3-14 by rotating that quadrilateral 180° around the centers of each of its edges (cf. Chapter I). Note that the sum of the angles of a quadrilateral is 360°, and that each of the angles of the tessellating quadrilateral is represented at each vertex as well. Therefore the sum of the angles around each vertex is also 360°, thus satisfying a condition for tessellating the plane.

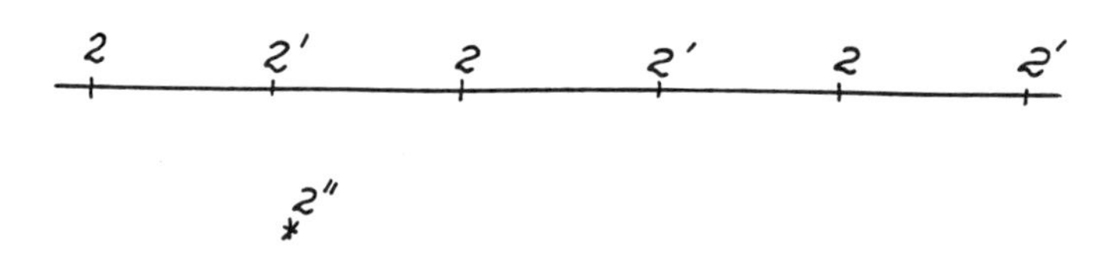

Figure 3-11 ***A third two-fold rotocenter not collinear with 2 and 2′***

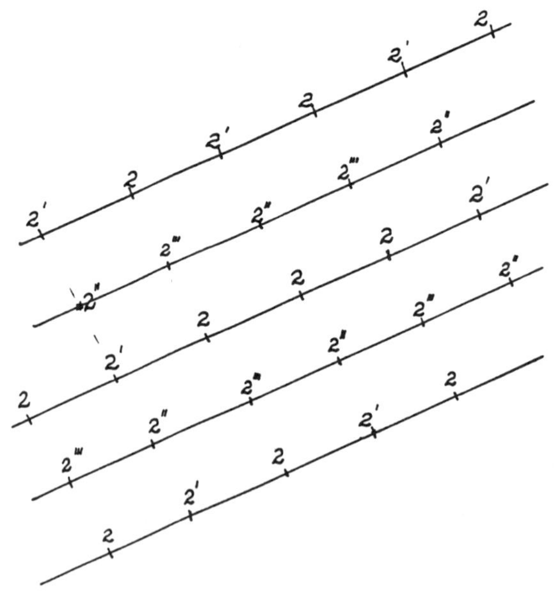

Figure 3-12 ***Four distinct sets of two-fold rotocenters***

Q: *DRAW THE DIAGONALS OF ANY QUADRILATERAL WHOSE EDGES HAVE TWO-FOLD ROTATIONAL SYMMETRY.*

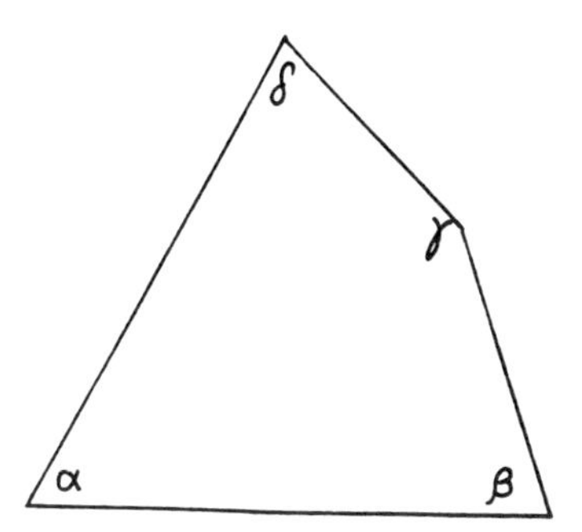

Figure 3-13 ***Quadrilateral having straight edges***

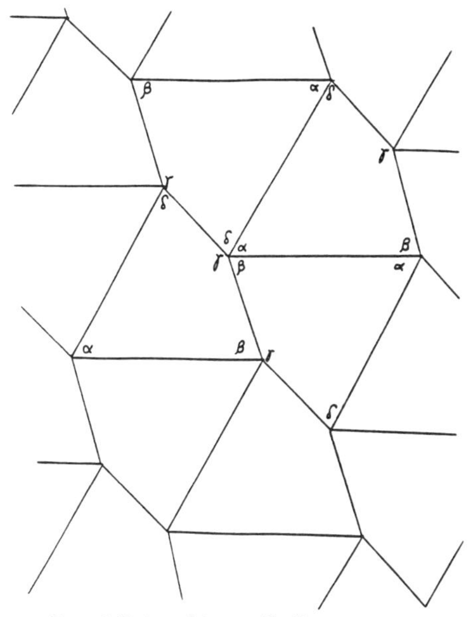

Figure 3-14 ***Quadrilateral tessellation***

Four lines joining the centers of rotational symmetry will each be parallel to one or the other diagonal, so that:

Theorem 3-1: If the edges of a quadrilateral have two-fold rotational symmetry, their centers of symmetry lie at the vertices of a parallelogram.

Figure 3-15 shows a tessellation generated from a module which is not convex, i.e., which has an internal angle greater than 180°. In Chapter I we claimed that it was not intuitively obvious that a tile of such form would tessellate the plane. We have now discovered by analysis that, indeed, the tessellation is possible.

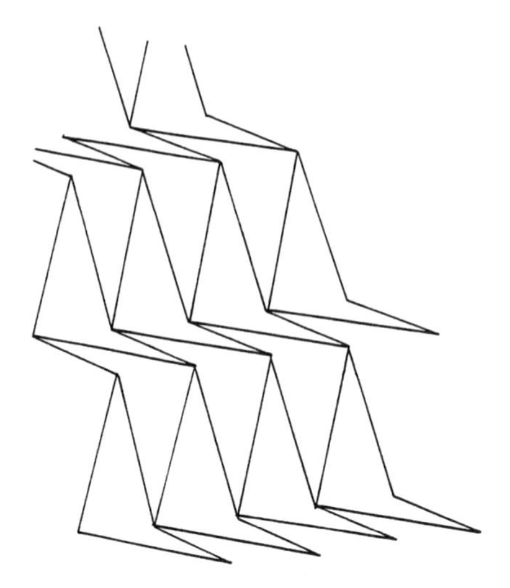

Figure 3-15 ***Tessellation by a concave straight-edged quadrilateral***

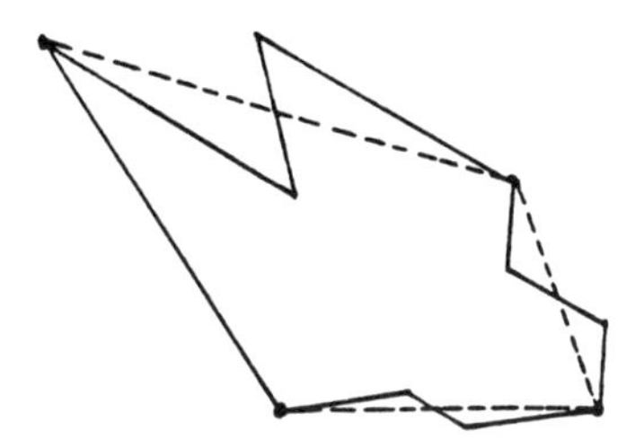

Figure 3-16 ***Curved-edged quadrilateral***

Figure 3-16 shows a quadrilateral constructed from a general straight-edged quadrilateral by scalloping its edges as illustrated: Each edge maintains its two-fold rotational symmetry. Rotating this curved-edged quadrilateral 180° around the center of each of its edges generates the tessellation of Figure 3-17. Three of the four edges in this example are composed of several straight-line segments, but the vertices of the original quadrilateral are the only vertices in the transformed tessellation. Only those points where more than two straight-line segments meet (in this instance four) are considered vertices of the tessellation. In this tessellation adjacent quadrilaterals (those sharing edges) are related by two-fold rotational symmetry, whereas those which share vertices only are related by translational symmetry. There is a marked analogy in this respect with the triangle tessellation of Figure 3-7.

Q: *PLACE A PIECE OF TRACING PAPER OVER FIGURE 3-17 AND INDICATE ALL ROTOCENTERS, DISTINGUISHING DISTINCT ROTOCENTERS BY MEANS OF PRIMES.*

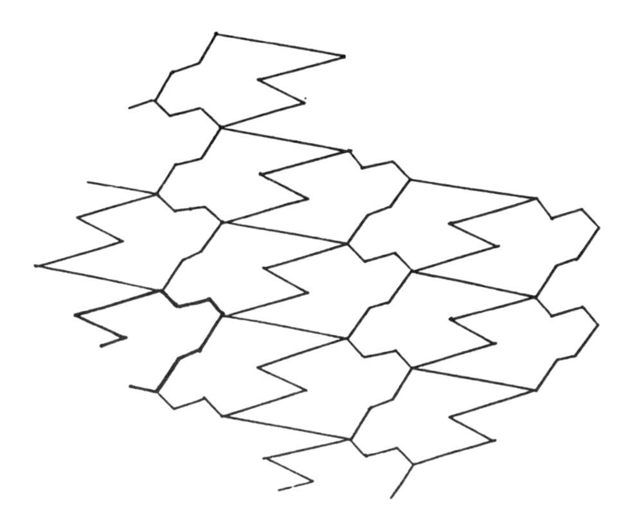

Figure 3-17 ***Tessellation by a curved-edged quadrilateral***

Once more the rotocenters lie equally spaced along equally spaced parallel rows.

There are a number of general conclusions to be drawn from these examples. In the first place, tessellations having translational symmetry require tiles whose opposite sides are pairwise parallel and congruent; triangles would therefore be excluded. Secondly, the quadrilateral tessellations of Figures 3-14, 3-15 and 3-17 have rotational as well as translational symmetry. Furthermore, translational symmetry, even when not the sole symmetry of the pattern, automatically appears as the result of the coexistence of two two-fold rotocenters. In the fourth place, three two-fold rotocenters not all on the same straight line imply a fourth two-fold rotocenter. The final generalization we may make is that any triangle and any quadrilateral, whether having straight or curved edges, will tessellate the plane as long as its edges have two-fold rotational symmetry.

One of the triangular tiles of Figure 3-3 is reproduced in the tessellation of Figure 3-18; the edges of this tile do not possess rotational symmetry. According to our definition, this tessellation is not triangular, however, but quadrilateral, for in traversing the circumference of one tile *in the tessellation*, one encounters *four*, not three vertices and edges: A vertex divides the arched top edge of the original triangular tile into two edges in the tessellation. This example shows that beside the curved-edge parallelogram tessellation and the quadrilateral tessellation having two-fold rotational symmetry there may be other, special ways of tessellating the plane. Some of these will be considered in subsequent chapters; for an exhaustive mathematical treatment the reader is referred to some of the references.

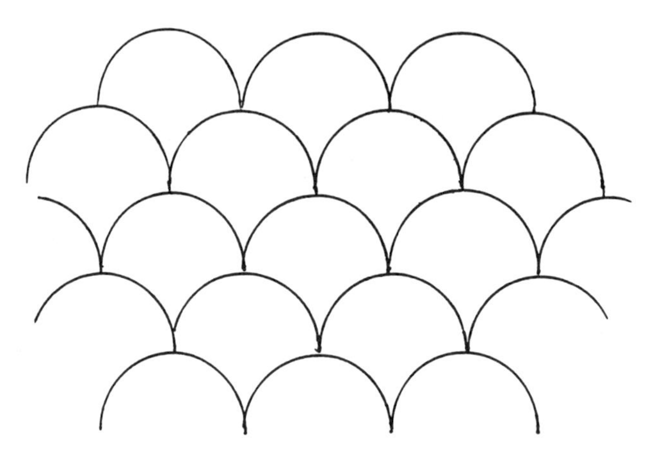

Figure 3-18 ***Quadrilateral tessellation generated from a triangle***

Q: *WHAT FRACTION OF THE AREA OF A CONVEX QUADRILATERAL WHOSE EDGES HAVE TWO-FOLD ROTATIONAL SYMMETRY LIES INSIDE THE PARALLELOGRAM WHOSE VERTICES ARE LOCATED ON THE CENTERS OF THE EDGES? (cf. Theorem 3-1)*

Figure 3-19 shows a quadrilateral PQRS and the parallelogram ABCD whose vertices lie at the midpoints of the edges of PQRS. The diagonals PR and QS are drawn. Since the length of the line PD is half that of PS, the area of the triangle PAD equals one quarter of that of triangle PQS (cf. Chapter II). The area of triangle RCB analogously equals one quarter of that of triangle RSQ. Therefore the combined areas occupied by triangles PAD and RCB equal one quarter the total area of quadrilateral PQRS. Analogously, the areas of triangles QAB and SCD add up to one quarter the area of quadrilateral PQRS, so that one half of the area of quadrilateral PQRS lies outside, and the other half inside parallelogram ABCD.

Let us compare a quadrilateral which has curved edges having two-fold rotational symmetry with a quadrilateral which has straight edges but the same vertices as the curved-edged quadrilateral. We then observe that both quadrilaterals have the same area, for the curved-edge quadrilateral may be obtained from the straight-edged one by redistributing, but not by adding or removing portions. Thus we conclude:

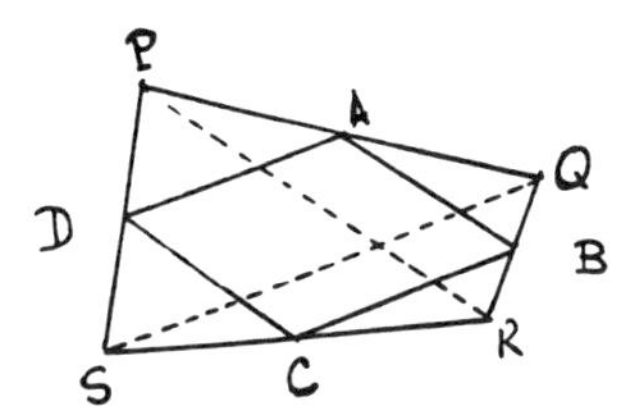

Figure 3-19 ***A quadrilateral and an inscribed parallelogram***

Theorem 3-2: The area of the parallelogram whose vertices are at the centers of two-fold rotational symmetry of the (generally curved) edges of a general quadrilateral equals exactly one half of the area of that quadrilateral.

There is a decided esthetic satisfaction in the discovery that a form as free as the general quadrilateral, whose only restriction is the two-fold rotational symmetry of its edges, should have inscribed in it the much more regular form of a parallelogram which occupies precisely one half of the area of the quadrilateral.

NOTES

[1] Chorbachi, Wasma'a K., and Arthur L. Loeb: *An Islamic Pentagonal Seal (From Scientific Manuscripts of the Geometry of Design)* in *Five-fold Symmetry*, ed., Hargittai, World Science, Singapore (1992), 283-303.

[2] Albairn, K., J. Miall Smith, S. Steele and D. Walker: *The Language of Pattern,* Thames & Hudson, London (1974).

[3] Bourgoin, J.: *Arabic Geometrical Pattern and Design*, Dover, New York (1973).
Critchlow, Keith: *Islamic Patterns,* Schocken, New York (1976).
El-Said, Issam, and Ayse Parman: *Geometric Concepts in Islamic Art*, World of Islam, London (1976).
Hoag, John: *Islamic Architecture*, Rizzoli, New York (1987).

[4] Loeb, A. L.: *Polyhedra: Surface or Volume?* in *Shaping Space*, M. Senechal and G. Fleck, eds. *Design Science Collection*, A. L. Loeb, series ed., Birkhäuser, Boston (1987).

[5] Grünbaum, Branko, and G. C. Shephard: *Tilings and Patterns*, Freeman, New York (1987).
Heesch, Heinrich: *Reguläres Parkettierungsproblem*, West Deutscher Verlag, Cologne (1968).
Schattschneider, Doris: *Tiling the Plane with Congruent Pentagons,* Math Magazine, **51**, 29–44 (1980).

IV

The Postulate of Closest Approach

In the previous chapter we noted that the coexistence of two distinct two-fold rotocenters implies an infinite row of equally spaced two-fold rotocenters (Figure 4-1). Furthermore, we generated a two-dimensional array of four distinct types of two-fold rotocenters by postulating a two-fold rotocenter off the row of two-fold rotocenters already generated (Figure 3-8). The question which we are to address here is what would happen if we placed a third distinct two-fold rotocenter *on* the line joining the equally spaced rotocenters.

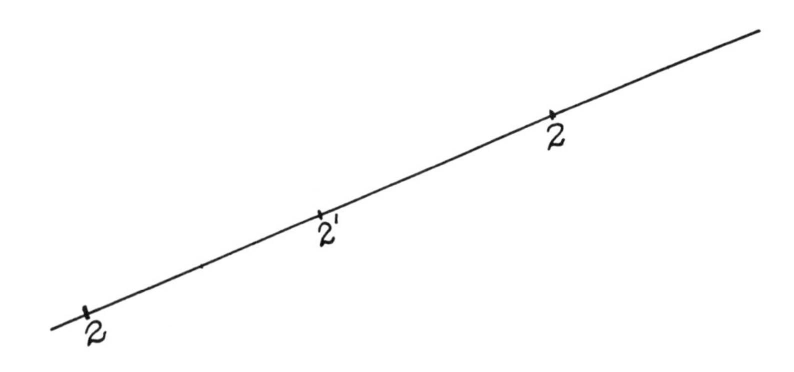

Figure 4-1 ***Equally spaced two-fold rotocenters***

To begin, let us place a two-fold rotocenter halfway between an adjacent pair 2 and 2′ and label it 2″, as it is not to be equivalent to either 2 or 2′ (Figure 4-2).

Q: *WHAT DOES 2″ IMPLY ABOUT 2 AND 2′?*

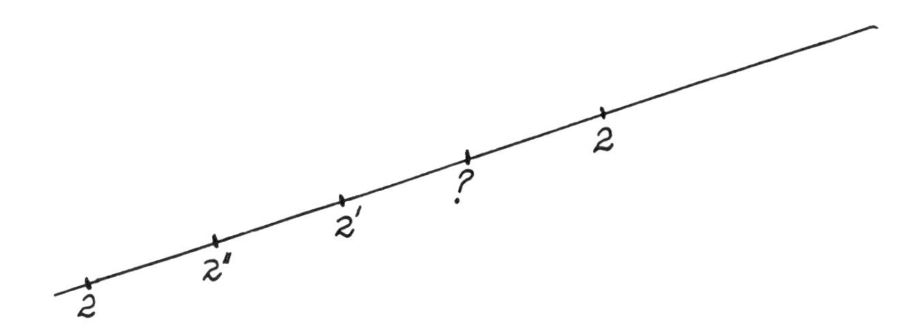

Figure 4-2 ***Three distinct collinear rotocenters***

The existence of 2″ implies that as a result of a two-fold rotocenter half-way between them, 2 and 2′ would be equivalent, with the result that the prime should be dropped from 2′ or added to 2. In either case there would be only *two* distinct kinds of rotocenters, 2 (or 2′) and 2″.

Next suppose that we had placed a 2″ twice as close to a 2 as to an adjacent 2′.

PLACE A PIECE OF TRACING PAPER OVER FIGURE 4-1 AND LOCATE A 2, A 2′ AND A 2″ AS INDICATED. LOCATE ALL IMPLIED ROTOCENTERS.

The result would have to be an additional 2 halfway between 2′ and 2″ (Figure 4-3). That newly generated 2 would imply, however, that the 2″ would need to be equivalent to 2′, so that a prime should be dropped from 2″ or added to 2′. In any case, an alternating row of equally spaced two-fold rotocenters would be generated, the spacing being one third as large as that of the original row.

From these examples it would appear that any attempt to place an additional two-fold rotocenter on a line defined by the coexistence of two distinct two-fold rotocenters will result in a new row of alternatively distinct rotocenters whose spacing is less than that of the original row. However, any attempt to make this third rotocenter distinct from the other two resulted in making two of the three "distinct" sets mutually equivalent, so that no more than two distinct sets of two-fold rotocenters coexisted on a straight line.

This argument may be generalized as follows for those who appreciate such mathematical niceties, and for those who still need to be convinced. Others may skip the following paragraphs within the square brackets without any loss of understanding of the general principle.

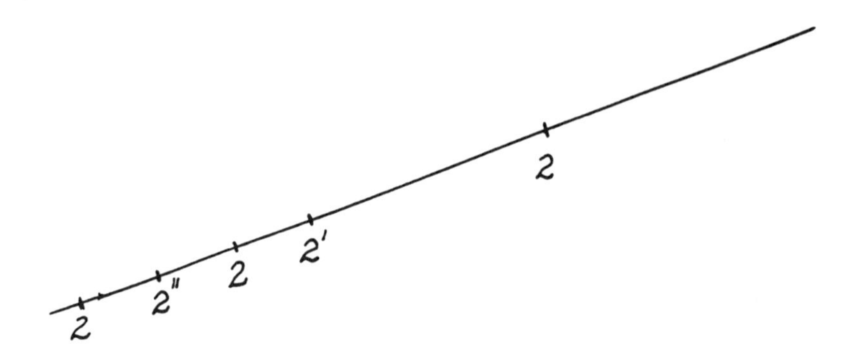

Figure 4-3 ***Implication of a third rotocenter collinear with 2 and 2'***

[If we call the distance between two adjacent two-fold rotocenters *unity*, then we might place a $2''$ at a distance $1/i$ from a 2, where i is a positive integer. The implied result would be 2-s at distances $2/i$, $4/i$, $6/i$, etc., generally at $2j/i$, where j is also a positive integer. There would be $2''$-s at distances $(2j-1)/i$. Accordingly, we would find at distance unity from the original 2 another 2 if i is even, but a $2''$ if i is odd. In point of fact, we postulated that there is a $2'$ at distance unity, for that is how we defined the unit distance. Therefore we should once more either add or drop a prime at $2'$, so that once more our attempt to introduce a third distinct set of two-fold rotocenters is foiled.

We could, of course, have placed $2''$ at a distance k/i from 2, where k is yet another positive integer; without loss of generality we shall restrict ourselves to values of k less than $\frac{1}{2}i$, so that $2''$ is closer to 2 than to $2'$. A new 2 is then implied at distance $2k/i$ from 2, which in turn generates a $2''$ at $3k/i$. There will be 2-s at $2jk/i$, $2''$s at $(2j-1)k/i$. In this case we do not necessarily find either a 2 or a $2''$ at distance unity; if not, then there will necessarily be both a 2 and a $2''$ at a distance less than k/i from unity (cf. Figure 4-4), the $2'$ at unity lying between a 2 at a distance less than unity and a $2''$ at a distance greater than unity from the original 2.

For instance, let us suppose that we had placed a $2''$ at distance $32/231$. Eventually, there would have been a $2''$ at $224/231$ and a 2 at $256/231$. The result would have been three distinct two-fold rotocenters, at respective distances $224/231$, $231/231$ (the original $2'$) and $256/231$ from the original 2. Where does that put us? To quote Anna Russell at the conclusion of her epic *Ring Cycle*, exactly where we started! Only the scale has changed, but there is no reason why we should not rescale ("zoom in") so that the interval between $224/231$ and $256/231$ becomes a new interval from 0 to 1. Then, after placing our new $2''$ at 0, the original $2'$ at $7/32$, and the new

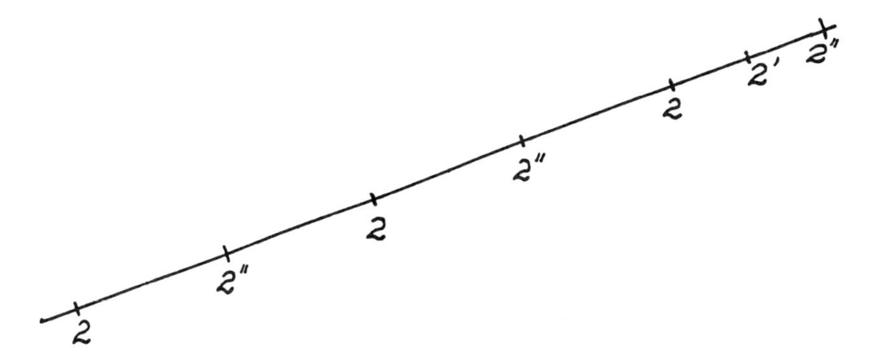

Figure 4-4 ***Implied rotocenters***

2 at 1, the coexistence of a $2''$ at 0 and a $2'$ at $7/32$ will imply a $2''$ at $14/32$, a $2'$ at $21/32$, a $2''$ at $28/32$, and a $2'$ at $35/32$. Once more, we are back where we started, this time with a $2''$ at $28/32$, a 2 at $32/32$, and a $2'$ at $35/32$. Scaling once more, we transform this interval to a $2''$ at 0, a 2 at $4/7$, and a $2'$ at 1. Because $4/7$ is greater than $\frac{1}{2}$, we should reverse the interval, placing $2'$ at 0, 2 at $3/7$, and $2''$ at 1. The coexistence of $2'$ at 0 and 2 at $3/7$ implies another $2'$ at $6/7$. The latter together with $2''$ at 1 implies a $2''$ at $5/7$, hence a $2'$ at $4/7$, and a $2''$ at $3/7$. Since there already was a 2 at $4/7$, 2 and $2''$ must be equivalent, so that we have once more found that there cannot be more than two distinct two-fold rotocenters on a straight line.

This same argument may be generalized as follows. Returning to our $2''$ at a general position k/i, we shall eventually find either a 2 at a position lk/i and a $2''$ at $(l+1)k/i$ or a $2''$ at lk/i and a 2 at $(l+1)k/i$, where $i/k-1 < l < i/k$. These inequalities assure that the original $2'$ lies between this pair of 2 and $2''$. In either case, we have three mutually distinct two-fold rotocenters in an interval which we may scale so that these three lie respectively at locations 0, $i/k-1$ and 1. Defining $m \equiv i - kl$, we then conclude that the original postulate of three distinct two-fold rotocenters respectively at 0, at k/i and at 1 will lead, with a scale change, to three distinct rotocenters respectively at 0, m/k and 1, where $0 < m < k$. In turn, the same argument will imply a trio at 0, n/m and 1, where $0 < n < m$. Continuing, we shall eventually find a trio at simple fractional distances such as 0, $1/3$, 1, for which, as we have already demonstrated, it is futile to attempt to create more than two distinct sets of two-fold rotocenters on a straight line.]

We conclude that two distinct two-fold rotocenters coexisting in a plane separated by a distance d preclude the existence of a third two-fold rotocenter distinct from both of these two, collinear with these two, and located at a distance fd from either, where f is a *rational* fraction. [This fraction f is our initial ratio k/i: We made extensive use of the fact that k and i are integers in generating a list of successive integers $k > l > m > n \ldots$, which would eventually terminate at a small integer.] If we had placed the original $2''$ at an *irrational* location between 0 and 1, then trios would continue to be generated, each set more closely spaced then the previous one, until our limit of resolution would be exceeded.

In principle, there is nothing wrong with having rotocenters crowding arbitrarily closely together. In practice, however, we are dealing with modules such as tiles in which rotocenters could not be so close together. Although there is a branch of mathematics which deals with such continuity of rotocenters (the theory of Lie groups), we shall exclude them from consideration here by a postulate which limits us to discrete structures:

POSTULATE OF CLOSEST APPROACH: For any pattern which we shall consider there exists a finite distance such that no two rotocenters may be closer to each other than that distance.

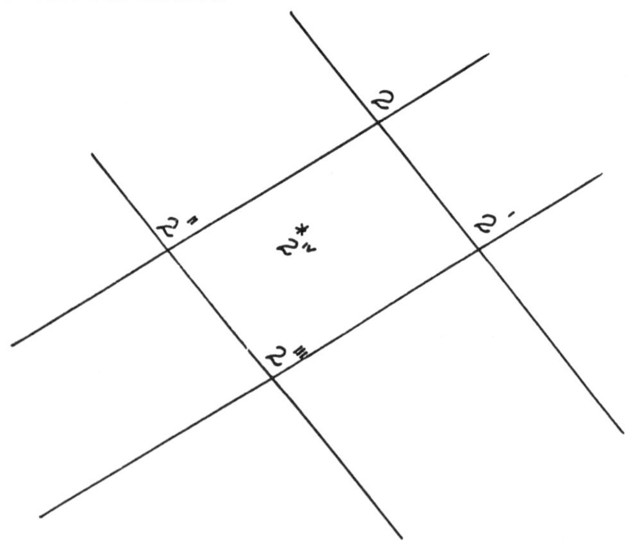

Figure 4-5 (a) ***Four distinct two-fold rotocenters in a plane, (a): With fifth two-fold rotocenter***

The postulate of closest approach eliminates the only possibility of having more than two distinct two-fold rotocenters on a straight line. It also eliminates placing a two-fold rotocenter distinct from the four distinct sets on any straight line joining any pair of rotocenters in Figure 4-5. In

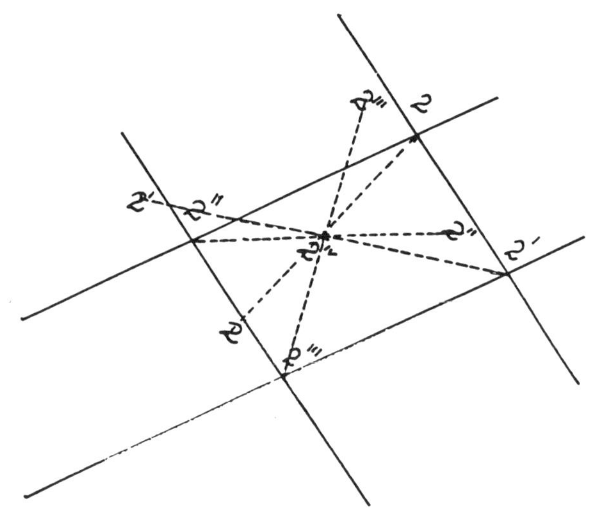

Figure 4-5 (b) ***Array of rotocenters implied***

Figure 4-5 (a) we show four distinct rotocenters, 2, $2'$, $2''$ and $2'''$, at the vertices of a parallelogram; we shall assume that these are all as close to each other as the postulate of closest approach allows. Somewhere inside this parallelogram we have placed a fifth two-fold rotocenter, 2^{iv}, distinct from the other four, and not lying on a straight line joining any of the other four. Just as in the case of our attempts to place additional distinct two-fold rotocenters on a straight line joining two distinct two-fold rotocenters, we shall also find that 2^{iv} will interact with each of 2, $2'$, $2''$ and $2'''$ to generate additional rotocenters equivalent respectively to 2, $2'$, $2''$ and $2'''$. When 2^{iv} interacts with these newly generated rotocenters, additional centers will be generated, so that eventually either the density of rotocenters will increase indefinitely, in violation of the postulate of closest approach, or the new center 2^{iv} will need to be equivalent to one of the other rotocenters (Figure 4-5 (b)). Thus we see that as long as the postulate of closest approach keeps us in the realm of discrete structures, no more than four distinct two-fold rotocenters may coexist in a plane. Figure 4-6 shows the four distinct rotocenters permitted in the plane.

A region of the plane whose vertices represent each of the distinct rotocenters in a plane, but which does not contain any rotocenter internally or elsewhere on its boundary, is called a *mesh*. The parallelogram $22'2''2'''$ discussed in the previous paragraph is a mesh.

Q: *USING THIS DEFINITION OF A MESH, COULD YOU FIND A DIFFERENTLY SHAPED MESH FOR THE ARRAY OF ROTOCENTERS IN FIGURE 4-6?*

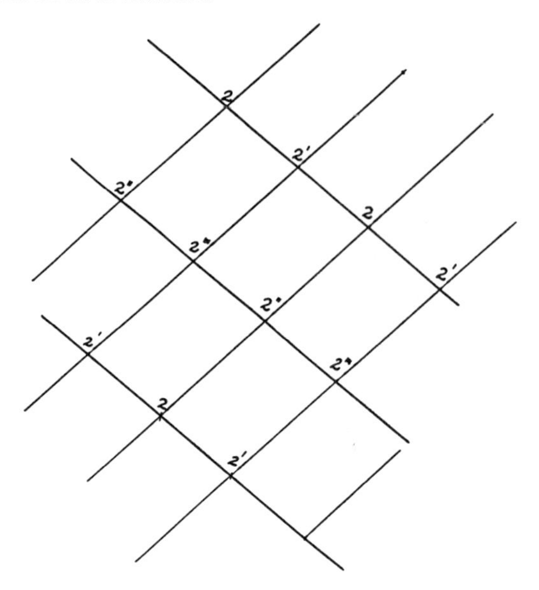

Figure 4-6 ***Array of two-fold rotocenters***

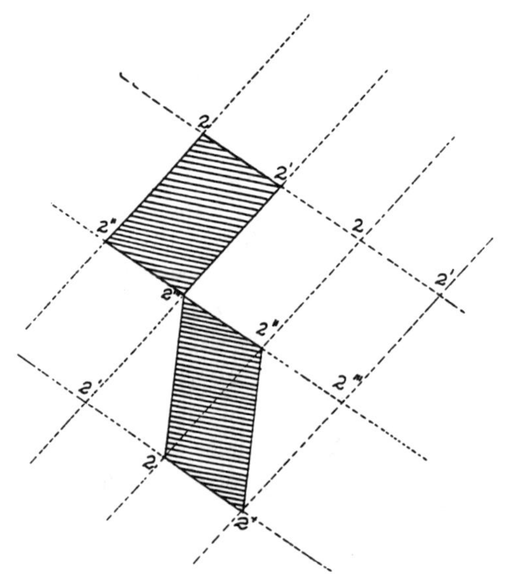

Figure 4-7 ***Several choices of meshes***

A mesh is not necessarily uniquely determined: In Figure 4-7 we show two different choices of meshes. What these meshes do have in common, however, is their area. In Figure 4-8 four meshes are superimposed upon the quadrilateral tiling pattern of Figure 3-17, which has two-fold rotational symmetry. Each of the vertices of the mesh lies, of course, at the midpoints

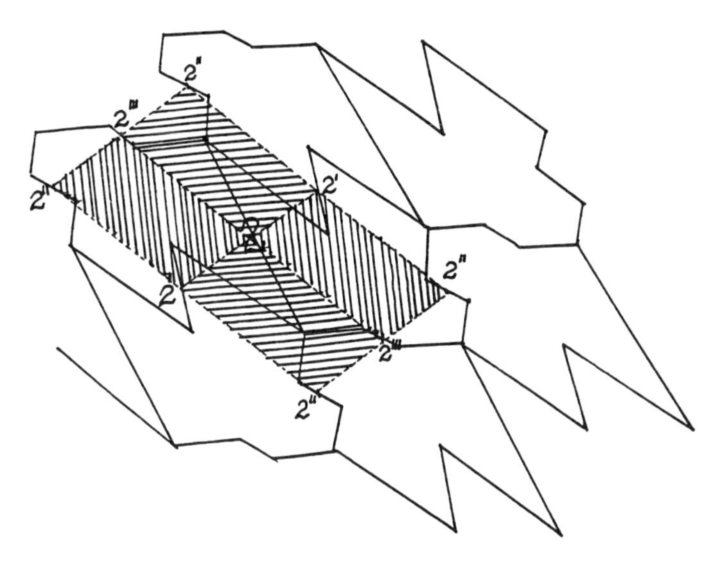

Figure 4-8 ***Comparison of tile and mesh in a tessellation***

of edges of a tile, and it is seen that the area of the title is just twice that of the mesh (cf. Theorem 3-2). To re-phrase this observation: Two meshes are needed to contain all the information contained in one tile. Note also that *four* meshes meet at each *two*-fold rotocenter.

V

The Coexistence of Rotocenters

A bicycle wheel appears to have rotational symmetry: When it is rotated through some angles smaller than 360°, it appears to be in the identical position to the one initially occupied, for the wheel is made up of a number of segments which are mutually congruent to each other. Actually, in practice there is a single valve, used to inflate the tire, which will mark a unique position on the wheel. Conceptually we may ignore the valve, and then determine the smallest angle through which the wheel must turn before it appears to assume its original position. If that angle, expressed in radians, is found to equal θ, then it will appear to assume that position $2\pi/\theta$ times during a complete rotation. The wheel will then be found to have $(2\pi/\theta)$-fold rotational symmetry; at the center of the hub there will be a $(2\pi/\theta)$-fold rotocenter.

At the center of a square we find a four-fold rotocenter; a three-fold rotocenter is at the center of an equilateral triangle, while the diagonals of a regular hexagon intersect at a six-fold rotocenter. The diagonals of the parallelogram intersect at a two-fold rotocenter. In Figure 5-1 we have juxtaposed a square and a parallelogram.

Q: *COULD WE COMBINE THESE TWO IN A PATTERN SO THAT EACH CONSERVES ITS OWN ROTATIONAL SYMMETRY?*

To resolve this question it is necessary to realize that a rotocenter applies to the entire universe. This means that the entire universe is divided into mutually congruent wedges which meet at the rotocenter. Any point a positive distance from a center of k-fold rotational symmetry, no matter how far removed from the rotocenter, will have $(k-1)$ equivalent points all lying at the same distance from the rotocenter. There is no such thing as "local" symmetry: Symmetry applies to the entire pattern, not just to a restricted region of a pattern.

We want to generate a pattern made up of squares and parallelograms which are congruent to the first pair in Figure 5-1 such that the overall

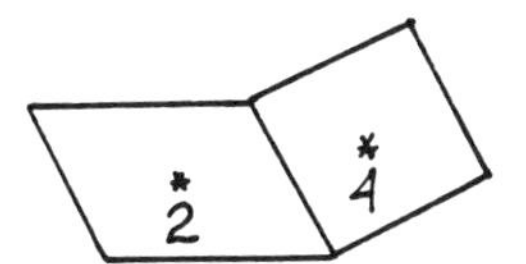

Figure 5-1 ***Combining a two-fold and a four-fold rotocenter***

pattern has a two-fold rotocenter at the center of each parallelogram and a four-fold rotocenter at the center of each square. (We do not need to tile the plane with these polygons: There will generally be space between them.) This condition requires that each parallelogram must be adjacent to two squares, and that four parallelograms must surround each square (Figure 5-2 (a)).

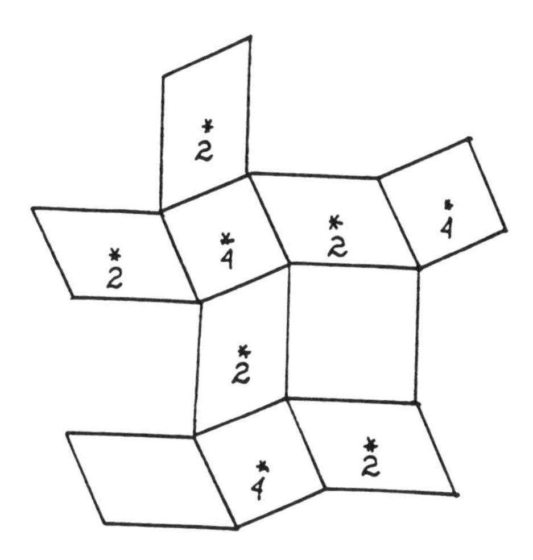

Figure 5-2 (a) ***Combining a four-fold and a two-fold rotocenter***

Q: *HOW MANY TILES DO WE NEED TO GENERATE A ROTOCENTER IN EACH TILE?*

The pattern so generated cannot be bounded, for any square or parallelogram at a boundary would not be symmetrically surrounded. A crystal diffracts x-rays because of the symmetrical arrangement of its component atoms and ions. The diffraction pattern is the clue to the symmetry surrounding these components. Nevertheless, crystals are bounded: There are external as well as, usually, internal boundaries at which there are components not so symmetrically surrounded. The number of such less symmetrically surrounded components is very small compared to the number of symmetrically surrounded ones, however, so that an unbounded pattern like the one generated above may be studied as an idealized model for real patterns. The x-ray diffraction pattern will be primarily determined by the symmetrical arrangement as long as the boundaries are sufficiently far apart.

Figure 5-2 (a), accordingly, is a portion of an unbounded, periodically repeating pattern. The two-fold rotocenters are labeled "2"; the four-fold ones at the centers of the squares are labeled "4."

Q: *DO YOU FIND ANY ADDITIONAL ROTOCENTERS IN THIS PATTERN?*

In addition to these rotocenters, which were postulated in generating the pattern, there is a second set of four-fold rotocenters whose context is quite distinct from the first set of four-fold rotocenters; to distinguish these implied rotocenters from the first postulated set, we have labeled them with a prime: "4′" in Figure 5-2 (b). The symmetry of the pattern of Figure 5-2 (b) is designated 244′.

The reader is invited similarly to juxtapose an equilateral triangle and a parallelogram, with the requirement that each of these polygons must preserve its rotational symmetry in the context of the pattern so generated. There are several ways of juxtaposing these polygons: one is shown in Figure 5-3.

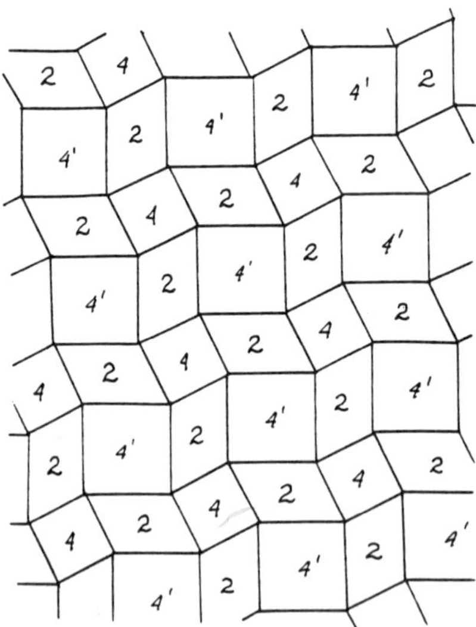

Figure 5-2 (b) ***Pattern having 244′ symmetry***

Q: *IS THERE ANY ROTOCENTER IN THE RESULTING PATTERN WHOSE VALUE ("FOLDNESS") IS DIFFERENT FROM 2 OR 3?*

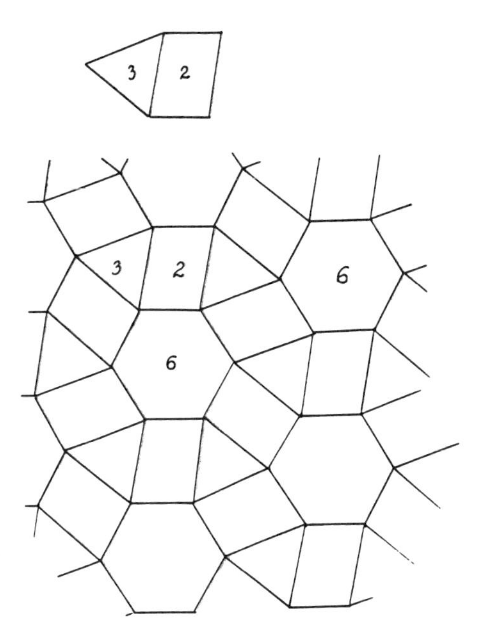

Figure 5-3 ***An equilateral triangle juxtaposed with a parallelogram***

In this case we observe implied *sixfold* rotocenters; the symmetry of this pattern is designated 236. Analogous designs should be attempted with pairs of polygons, each having rotational symmetry values 2, 3, 4, 5, 6, 7, 8, etc. The results should be entered in Table 5-1 (a); note that the lower left-hand triangle of this table would duplicate the upper right-hand one, hence may be omitted.

For combinations of two distinct rotocenters having the same symmetry values, e. g., two three-fold rotocenters, be sure to use two polygons which, although they may be geometrically similar (for instance, two squares or two equilateral triangles), are not mutually congruent. Also be sure to use primes to distinguish distinct rotocenters at the centers of these polygons. For the combination (2, 2′) we recall (Chapters III and IV) that a row of alternating two-fold rotocenters is generated: The implied symmetry is *translational*, to be denoted *tr* in Table 5-1 (b).

You will find that all combinations do not work out equally conveniently.

Q: *WHAT HAPPENS WITH THE COMBINATION 3 AND 4? HOW ABOUT 6 AND 6′?*

Table 5-1 (a): ***Symmetry Implied by Coexisting Rotocenters.***

	2	3	4	5	6	7	8...
2							
3	-						
4	-	-					
5	-	-	-				
6	-	-	-	-			
7	-	-	-	-	-		
8	-	-	-	-	-	-	

Table 5-1 (b): ***Symmetry Implied by Coexisting Rotocenters.***

	2	3	4	5	6	7	8...
2	*tr*	6	4		3		
3	-	3			2		
4	-	-	2				
5	-	-	-				
6	-	-	-	-			
7	-	-	-	-	-		
8	-	-	-	-	-	-	

For the combination of a three-fold and a four-fold rotocenter a problem arises because rotocenters appear to want to crowd closer together when more are generated. The tiles themselves seem to be too large to permit four triangles to fit around each square while simultaneously three squares have to fit around each triangle. Let us take a piece of paper and mark two points labeled respectively A_3 and B_4 (Figure 5-4 (a)). The existence of B_4 implies three additional rotocenters equivalent to A_3, of which only one, A_3', is shown in Figure 5-4 (b). Because B_4 is a four-fold rotocenter, the angle $A_3B_4A_3'$ equals ninety degrees.

A_3' implies rotocenters equivalent to B_4, of which one, B_4' is shown in Figure 5-4 (c); the angle $B_4A_3B_4$ equals 120°. As shown in Figure 5-5, A_3'' and B_4'' are similarly generated.

Q: *WHICH DISTANCE IS LARGER, A_3B_4 OR A_3B_4''?*

Note that the rotocenter B_4'' is closer to A_3 than is B_4! This means that the coexistence of a three-fold and a four-fold rotocenter at a given distance from each other imply the existence of a new four-fold rotocenter closer to the three-fold rotocenter than was the first four-fold rotocenter.

Figure 5-4 (a) *Two coexisting rotocenters*

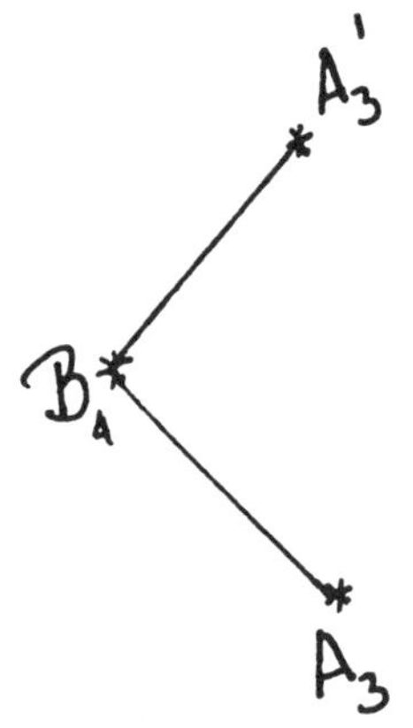

Figure 5-4 (b) *Two coexisting rotocenters together with a third, implied center*

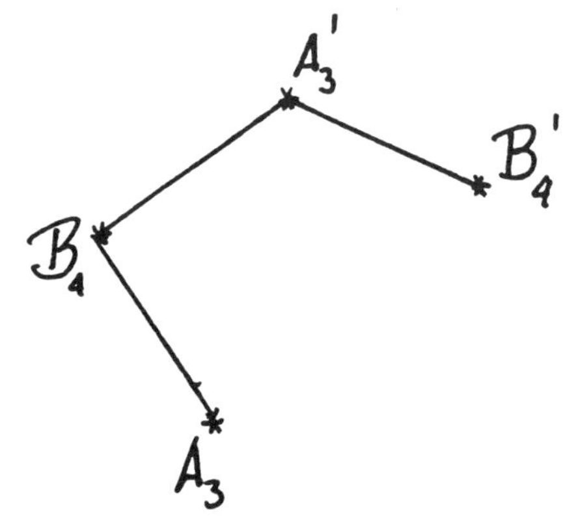

Figure 5-4 (c) *Continued generation of rotocenters*

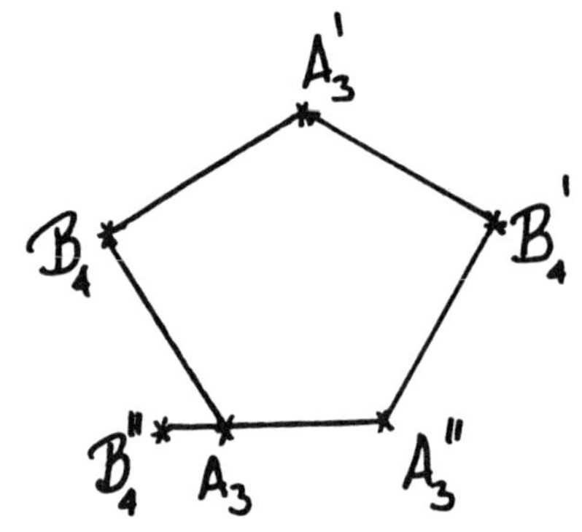

Figure 5-5 ***Going all the way around***

This same argument should then be applied to the new four-fold rotocenter, implying an additional four-fold rotocenter even closer to the original three-fold rotocenter. Continuing along this logically necessary argument, we will ultimately violate the postulate of closest approach, generating rotocenters separated by an arbitrarily small distance. We find that coexistence of a three-fold and a four-fold rotocenter in a plane implies a violation of the postulate of closest approach, and therefore cannot be permitted.

Q: *WHAT DO YOU CONCLUDE ABOUT THE COMBINATION (4, 6)?*

Let us consider next the combination (6, 6′) illustrated in Figure 5-6. Here A_6 and B_6 coexist at a distance r apart. A'_6 is generated $\pi/3$ radians clockwise from the A_6 on a circle, centered on B_6, having radius r. This A'_6, however, will generate a B'_6 exactly at the location of the original A_6, with the result that A_6 and B_6 are in point of fact equivalent. This result is at variance with our basic premise that there be two *distinct* six-fold rotocenters, so that we must conclude:

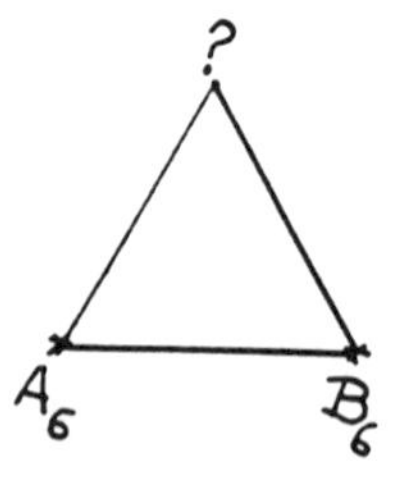

Figure 5-6 ***Two distinct six-fold rotocenters coexisting in a plane***

Theorem 5-1: All six-fold rotocenters coexisting in a plane are necessarily mutually equivalent.

Let us look at the problem in a more general manner. Take a k-fold rotocenter, A_k, and a l-fold rotocenter, B_l, separated from each other by a distance r. There will be k centers equivalent to B_l spaced by $2\pi/k$ radians on the circumference of a circle having radius r and center A_k. In turn, B_l is at the center of a circle having radius r; on the circumference of this circle there are l rotocenters equivalent to A_k spaced by $2\pi/l$ radians. Consider for the present only the one that is just $2\pi/l$ radians counter-clockwise from the original A_k (Figure 5-7). In turn, k rotocenters B_l are located at distance r from this latter A_k, of which we choose just $2\pi/k$ radians counter-clockwise from the one previously generated. Continuing in this fashion, we appear to generate a polygon, all of whose edgelengths equal r, whose angles are alternately $360°/l$ and $360°/k$. Certainly this is the case for the combinations (2,3), (2,4), (2,6), (3, 3′), (3,6), (4,4′): The implied rotocenters have symmetry values shown in Table 5-1 (b).

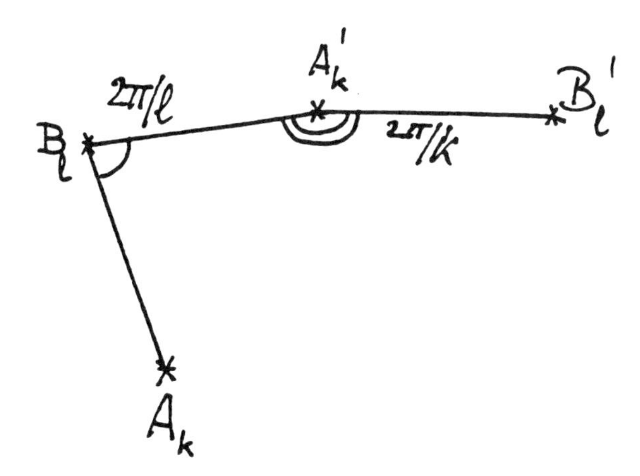

Figure 5-7 ***Two coexisting rotocenters together with an implied center***

This leaves for consideration all cases involving five-fold rotational symmetry and all symmetry values higher than six, since we have already eliminated the combinations (3,4), (4,6) and (6,6′).

Returning to Figure 5-7, let us label successively generated rotocenters A_k, B_l, A_k',B_l', A_k'', B_l'', A_k''', B_l''',.... Then bisect angles $A_kB_lA_k'$, $B_lA_k'B_l'$, $A_k'B_l'A_k''$, $B_l'A_k''B_l''$, etc. Let the first two bisectors intersect at a point C (Figure 5-8). The line segments A_kB_l, B_lA_k', $A_k'B_l'$, B_kA_k'', etc., all equal r

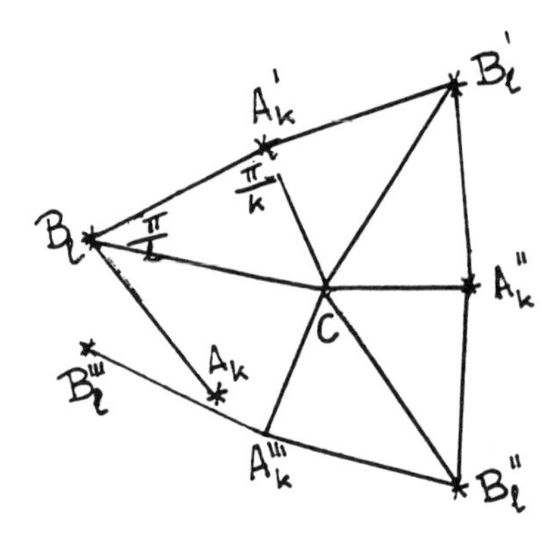

Figure 5-8 ***Intersecting bisectors***

in length. Angle A_kB_lC = angle $A'_kB_lC = \pi/l$, and angle B_1A_kC = angle $B_lA'_kC = \pi/k$. Clearly, the bisector at B'_l, making an angle π/l with line segment $A'_kB'_l$, will pass through C, as do all successive bisectors. Triangle A_kB_lC is congruent with triangles $A'_kB'_lC$, $A''_kB''_l$ C, $A'''_kB''_lC$, etc.: some of the latter can be brought into coincidence with the former only by flipping them over out of the plane.

$$\text{Angle} \quad A_kCB_l = \pi(1 - k^{-1} - 1^{-1})\,.$$

Q: *WHAT WOULD THIS ANGLE BE FOR THE EXAMPLE (6,6′)? WOULD SUCH TRIANGLES FIT AROUND C? IF SO, WHAT WOULD YOU CONCLUDE ABOUT A_6 AND B_6?*

Angle A_kCB_l is, for the example (6,6′), equal to $2\pi/3$; only three triangles will fit around C: A_6B_6C, $B_6A'_6C$, and $A'_6B'_6C$. As we have seen, that would imply $A_6 = B'_6$. Generally, then, we must have an *even* number of angles equal to $\pi(1 - k^{-1} - 1^{-1})$ fitting around point C:

$$2\pi m\,(1 - k^{-1} - 1^{-1}) = 2\pi\,, \quad \text{where } m \text{ is an integer.}$$

Hence:

$$k^{-1} + 1^{-1} + m^{-1} = 1\,. \tag{5-1}$$

If an *odd* number of these angles fit around C, then A_k and B_l can no longer be distinct, and if, on going around point C, we under- or overshoot by a fraction of that angle, we will violate the postulate of closest approach. On the other hand, if equation 5-1 is satisfied, then point C will be a rotocenter whose symmetry value equals m.

Theorem 5-2: The coexistence in a plane of a k-fold and an l-fold rotocenter is possible only if k and l satisfy the equation

$$k^{-1} + l^{-1} + m^{-1} = 1,$$

where k, l and m are integers, in which case an m-fold rotocenter is implied.

Q: *FIND m FOR THE CASE k=3, 1=4.*

If $k = 3$ and $l = 4$, m would have to be 12/5; if $k = 4$ and $l = 6$, m would have to be 12/7; neither of these are integers. Therefore the combinations (3,4) and (4,6) do not work.

If $k = 5$, then there are no integral values of l and m which will satisfy equation 5-1: an angle $\pi/5$ would not be able to conform to any other angle $\pi/1$ to achieve a closed polygon. Thus:

Theorem 5-3: No five-fold rotocenter may coexist in a plane with any other rotocenter.

VI

A Diophantine Equation and its Solutions

Our next task is to solve equation 5-1 in order to find all permissible combinations of rotocenters. Equation 5-1 is of a special kind called *diophantine*, after Diophantes of Alexandria, who is presumed to have discovered them. In general, all variables in such an equation are to be rational; in our case they are integers. Although in general one cannot solve a single equation in three variables, the restriction that the variables be integers limits us to a finite number of solutions.

Without loss of generality, we may assume that $k \leq l \leq m$, for equation 5-1 is symmetrical in k, l and m; that is to say, the equation is unaffected by interchanging any two of them. For $k = 1$, both l and m would have to be infinite. Of course, one-fold symmetry means invariance to a 360° rotation, which means no rotational symmetry at all. To understand infinite-fold symmetry, consider the pendulum of Figure 6-1. This pendulum, of length L, is to swing with an amplitude a. The length of the pendulum is variable: We shall move the suspension point O, keeping the amplitude a constant.

Q: *WHAT HAPPENS TO THE CURVE DESCRIBED BY THE BOB WHEN WE LENGTHEN THE PENDULUM KEEPING THE AMPLITUDE CONSTANT?*

The angle θ through which the pendulum swings will decrease as L increases. With increasing length L the circular arc described by bob of the pendulum will flatten out and more closely approach a straight line of length a.

Accordingly, rotation around the point O will become indistinguishable from translation as the radius L increases indefinitely while its amplitude remains constant; we may therefore regard translational symmetry as infinite-fold rotational symmetry around a rotocenter infinitely far away. The solution $(1\infty\infty')$ thus corresponds to translational symmetry in two independent directions and is exemplified by a tessellation of the plane with curved-edged parallelograms (cf. Figure 3-5).

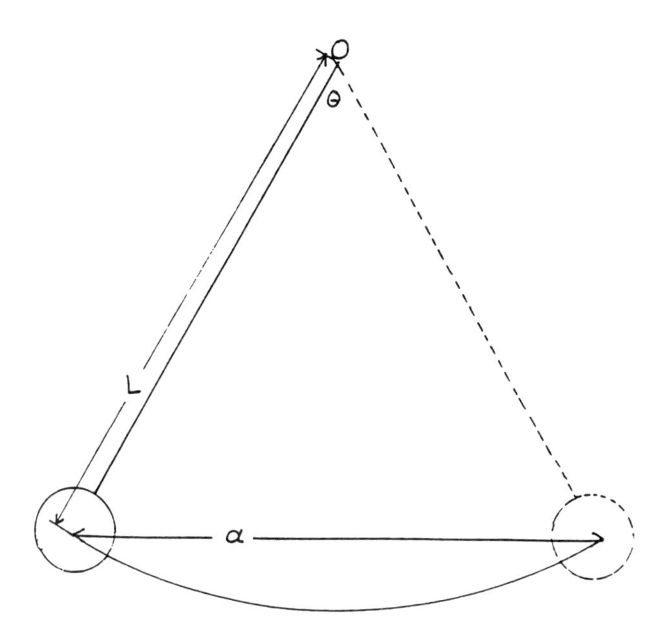

Figure 6-1 ***Pendulum***

When $k = 2$ and $l = 2$, $m = \infty$.

Q: *ACCEPTING THIS INTERPRETATION OF TRANSLATIONAL SYMMETRY AS INFINITE-FOLD ROTATIONAL SYMMETRY, COULD YOU THINK OF A PATTERN CORRESPONDING TO THE SOLUTION 22′∞ FOR THE DIOPHANTINE EQUATION?*

This solution corresponds to the alternatingly distinct two-fold rotocenters shown in Figure 3-10. It is periodic in one direction only, unlike Figure 3-5, which corresponds to periodicity in two independent directions. Patterns which repeat periodically in only one direction will be called *monoperiodic*, those repeating in two independent directions *diperiodic*. Frescoes are examples of monoperiodic patterns. We shall return to this solution presently, but we will first complete the solution of Equation 5-1.

When $k = 2$, there are also the possibilities $l = 3$, $m = 6$, and $l = 4$, $m = 4$. The former is exemplified by a hexagonal tessellation, in which the sixfold rotocenters are at the centers, the threefold ones at the vertices, and the twofold rotocenters at the centers of the edges of the tiles (Figure 6-2). The latter is exemplified by a square tessellation: One set of four-fold rotocenters is at the centers of the tiles, the other at their vertices; twofold rotocenters are again at the centers of the edges.

From equation 5-1 we further learn that when $k = 2$ and $l > 4$, m would have to be smaller than 4, but that would mean that $l > m$, which is counter to our agreement $k \leq l \leq m$. Therefore we have herewith exhausted all solutions having $k = 2$. For $k = 3$, l would need to be at least 3, and $l > 3$ would imply $m < 3$, hence $m < 1$, in violation of our initial agreement. Therefore $k = l = m = 3$ is the only solution having $k = 3$.

Q: *WOULD A TESSELLATION OF REGULAR TRIANGLES CORRESPOND TO SYMMETRY 33′3″?*

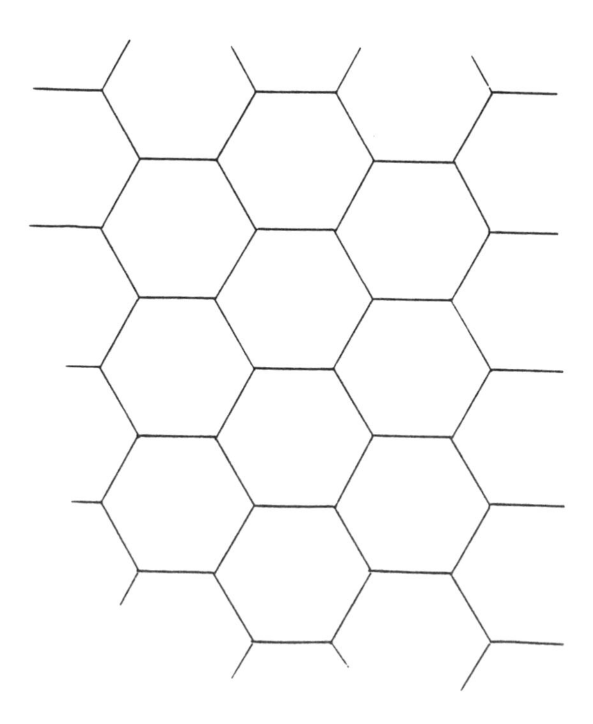

Figure 6-2 ***Hexagonal tessellation***

A tessellation of regular triangles corresponds to symmetry 236, as the vertices are six-fold rotocenters, three-fold rotocenters are at the centers of the triangles, and two-fold rotocenters are at the centers of the edges.

If k were to exceed 3, l and m would both need to be less than k, again in violation of the basic agreement. The exhaustive list of solutions of equation 5-1 is, accordingly as given in Table 6-1.

Table 6-1: ***Exhaustive Solution of the Diophantine Equation.***

k	l	m
1	∞	∞
2	2	∞
2	3	6
2	4	4
3	3	3

The table confirms what we found in Theorems 5-1 and 5-3, namely that a five-fold rotocenter precludes the existence of any other rotocenter in the same plane, and that all six-fold rotocenters in the same plane are necessarily equivalent to each other. The table also accounts for all entries in Table 5-1 which did or did not work; it also contains no number greater than 6, so that we may conclude:

Theorem 6-1: No rotocenter having symmetry value greater than six can coexist in the plane with any other rotocenter.

The notation and classification of two-dimensional symmetrical patterns developed in the context of crystallographic research is not necessarily most suitable for use in an art-historical or design context.[1] There are even differences between crystallographers and solid-state scientists in the manner in which they deal with symmetrical patterns.

Fundamental to *crystallographic* notation is the concept of the *lattice*, a collection of all points in a pattern related to each other by *translational* symmetry. This emphasis on translational symmetry is related to the translational symmetry of the x-ray beam used by crystallographers to determine the location of the crystal elements: Diffraction of the beam by the crystal is a direct result of this translational symmetry of the crystal. Solid state scientists, however, are more concerned with the symmetries of the fields, electrical, magnetic or quantum-mechanical, around each crystal element than with the absolute orientation of these fields.

We shall present here a notation based on the solutions of the diophantine equation, in which rotational symmetry is fundamental. Patterns having only a single rotocenter are denoted by the symmetry value of that rotocenter, k. Translational symmetry in a single direction is designated ∞. We noted that the coexistence of two rotocenters is limited by the diophantine equation, and that

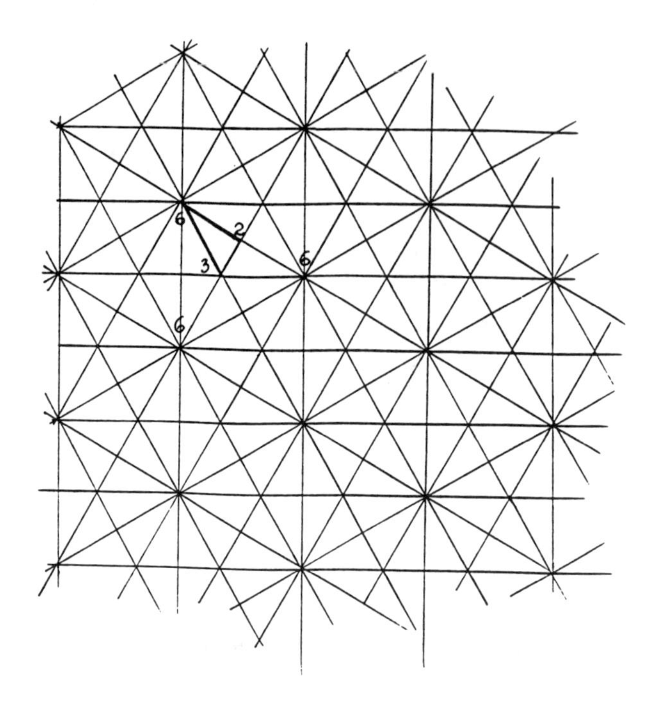

Figure 6-3 ***Meshes in the 236 system***

the permitted combinations imply a third rotocenter, resulting in the following five combinations: $1\infty\infty', 22'\infty, 236, 244', 33'3''$, where distinct rotocenters having the same symmetry value are distinguished by a prime: k and k'.

We have encountered patterns in which there are four distinct two-fold rotocenters; these are denoted $22'2''2'''$. We also noted that no more than four distinct two-fold rotocenters are permitted to coexist without violating the postulate of closest approach. To prove this upper limit to the number of distinct two-fold rotocenters in the plane, we used the concept of a *mesh*, a polygon whose vertices are rotocenters but which do not contain rotocenters. Figures 6-3, 6-4 and 6-5 show the meshes corresponding respectively to the rotocenter combinations 236, 244′ and 33′3′′′, and the way in which they tessellate the plane. Note that the number of meshes meeting at each rotocenter equals *twice* the symmetry value of that rotocenter. This means that no pair of points in mutually adjacent meshes can be related by rotational symmetry.

The question which we need to address now is whether more than three distinct rotocenters can coexist in a plane if at least one of these has symmetry value greater than 2. If three such rotocenters coexist, then they necessarily belong to one of the systems $22'\infty, 236, 244', 33'3''$, or $22'2''2'''$. In the last four

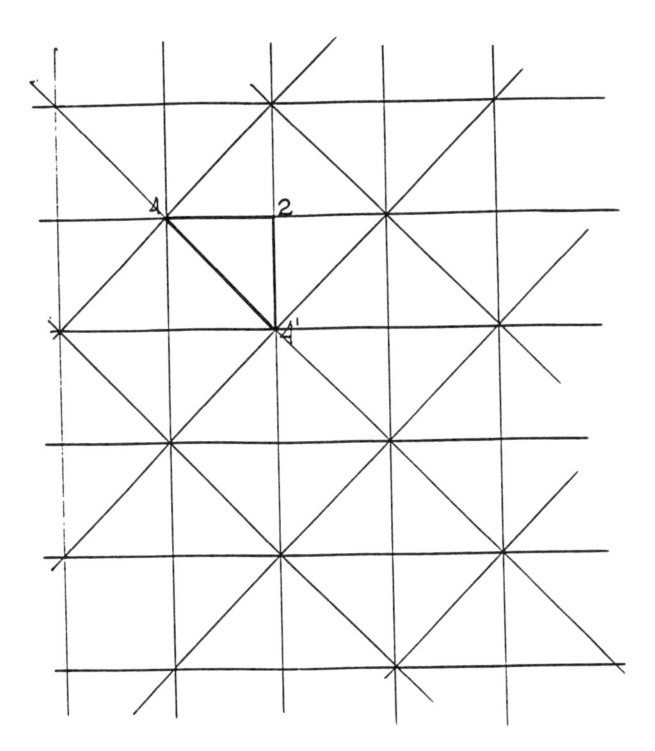

Figure 6-4 ***Meshes in the 244′ systems***

of these no additional rotocenters could be accommodated without violating the postulate of closest approach: In the case $22'\infty$, any rotocenter having symmetry value greater than 2 would interact with each of the 2 and $2'$ rotocenters to generate a pair of 236 or $244'$ patterns, whose coexistence would violate the postulate of closest approach. Thus all we can add to $22'\infty$ is a $2''$, which, as we have seen, implies $2'''$, generating the $22'2''2'''$ system.

The $22'2''2'''$ system differs fundamentally from the others in more than one respect. Not only does it have four distinct sets of rotocenters, hence quadrilateral meshes whereas the others have triangular meshes, but the shape of these quadrilateral meshes is not uniquely determined, whereas the triangular meshes all have their angles fixed. In the $22'2''2'''$ system, the mesh is a parallelogram whose angles may be chosen arbitrarily.

We have noted that the patterns in the system $22'\infty$ are monoperiodic. So are the patterns in the system ∞, having translational symmetry in one direction only. Both the system $1\infty\infty'$ and $22'2''2'''$ are diperiodic but involve

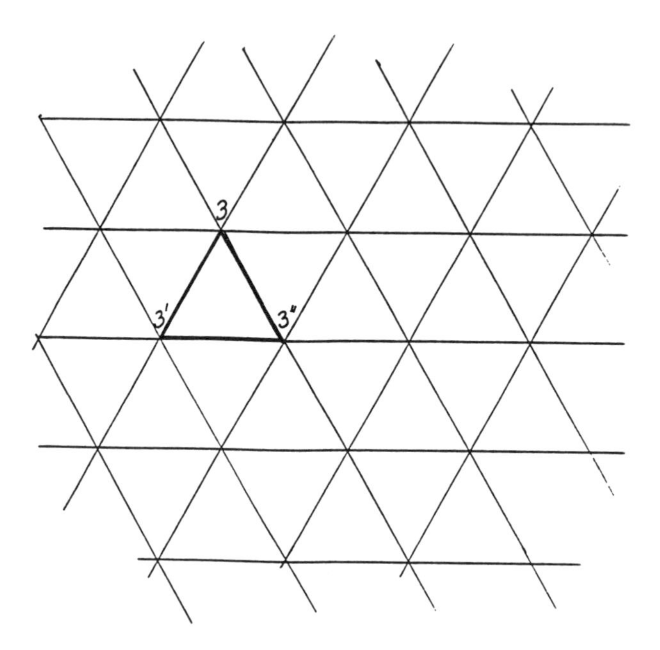

Figure 6-5 ***Meshes in the 33′3″ system***

an arbitrary angle: in $1\infty\infty'$ this is because there is translational symmetry in two arbitrary non-parallel directions, while in $22'2''2'''$ it is because of the arbitrary angle of the parallelogram mesh. It is almost as if in both cases the monoperiodic patterns are stacked on top of each other in arbitrary juxtaposition, just as the 234, 244′ and 33′3″ patterns can be stacked in various ways to generate three-dimensional patterns. We may therefore think of the $1\infty\infty'$ and $22'2''2'''$ systems as transitional between two-dimensional and three-dimensional patterns.

Summarizing (Table 6-2), we classify periodic patterns according to their rotational symmetry.

We approached symmetrical patterns from the perspective of tessellations of the plane. The meshes are particular tiles whose vertices are rotocenters. There are, however, many periodically repeating patterns which are not tessellations. Wallpapers are an example. Nevertheless, such patterns can always be subdivided into meshes (Figure 6-6 (a) and 6-6 (b)). A designer may place any design or motif in a mesh, remembering, however, that there should not be any rotocenter inside the mesh. As no pair of points in adjacent meshes are related by rotational symmetry, the designer is free to fill adjacent meshes with different designs, or to leave half of the meshes empty.

Table 6-2: ***Classification According to Rotational Symmetry.***

Symmetry Values	*Remarks*
$k = 1$	No rotational or translational symmetry.
$1 \leq k < \infty$	A single rotocenter having finite symmetry value.
∞	Translational symmetry in a single direction.
$22'\infty$	Alternating two-fold rotocenters along a straight line.
236	Meshes are right triangles having 30° and 60° angles.
244′	Meshes are right triangles having 45° angles.
33′ 3″	Meshes are equilateral triangles.
22′2″2‴	Meshes are parallelograms.

Figure 6-6 (a) was generated by placing a curve with one end on a three-fold rotocenter and then repeating the curve such as to preserve the symmetry of the rotocenter. A second, distinct, three-fold rotocenter is placed at some distance from the first rotocenter and its affiliated three identical curves. The entire pattern must then be repeated, and the curves extended until they just meet their replicas. Not surprisingly, a third set of rotocenters, distinct from the first two sets, now emerges. Note that a straight line through two distinct three-fold rotocenters of necessity also passes through a three-fold rotocenter distinct from both others.

Figure 6-6 (a) ***Design which has 33′ 3″ symmetry and two distinct types of tiles***

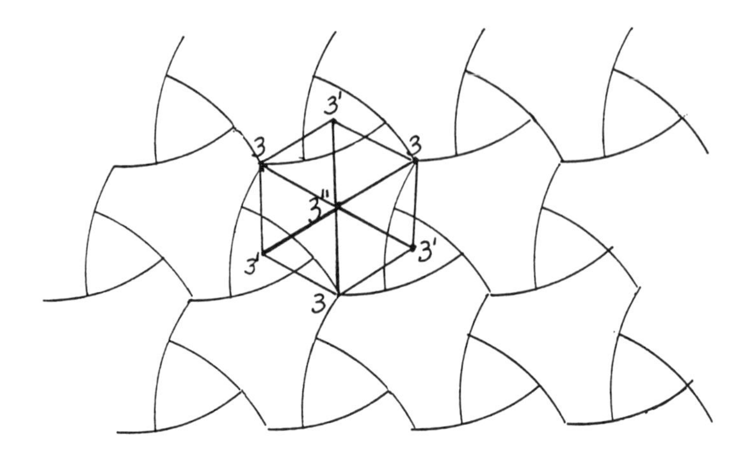

Figure 6-6 (b) ***Meshes in the design of Figure 6-6 (a)***

Six meshes for the pattern in Figure 6-6 (a) are shown in Figure 6-6 (b). Any pair of adjacent meshes contains all the information (Figure 6-6 (c)), which is repeated throughout the design. Note that most of the generating curve is contained continuously within the pair of adjacent meshes, but the bit clipped off by the mesh boundary reappears from an adjacent boundary to meet itself!

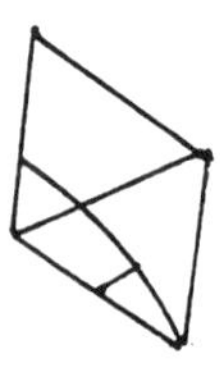

Figure 6-6c ***A pair of adjacent meshes in the design of Figure 6-6 (a)***

In both Figures 6-7 (a) and 6-7 (b) there are two- and four-fold rotocenters.

Q: *WHICH OF THESE TWO PATTERNS WOULD YOU CONSIDER TO BE MORE SYMMETRICAL?*

The difference in these two designs is subtle: In Figure 6-7 (a) lines intersect, while in Figure 6-7 (b) they interrupt each other. (We tend to interpret interruptions three-dimensionally, as in a weaving design.) In Figure 6-7 (a), one set of four-fold rotocenters occurs where the lines cross, the other in the spaces between the lines. Two-fold rotocenters lie *on* the lines, halfway between the crossings.

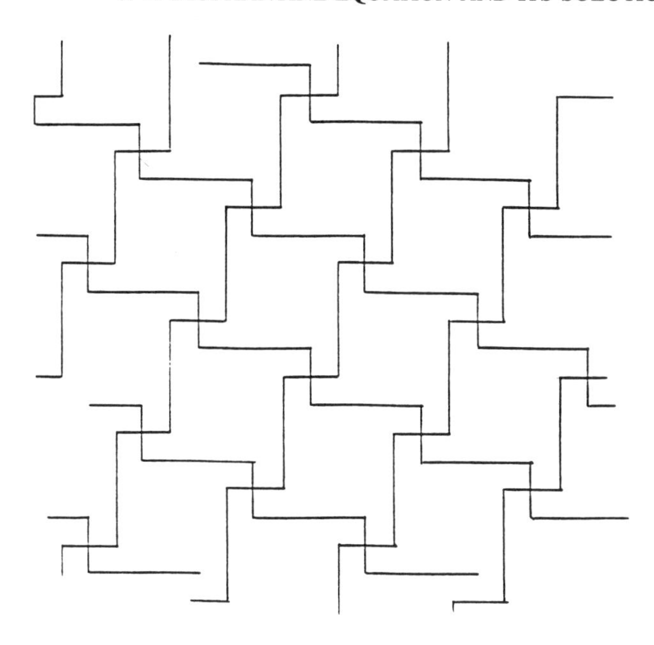

Figure 6-7 (a) ***Design having two- and four-fold rotocenters***

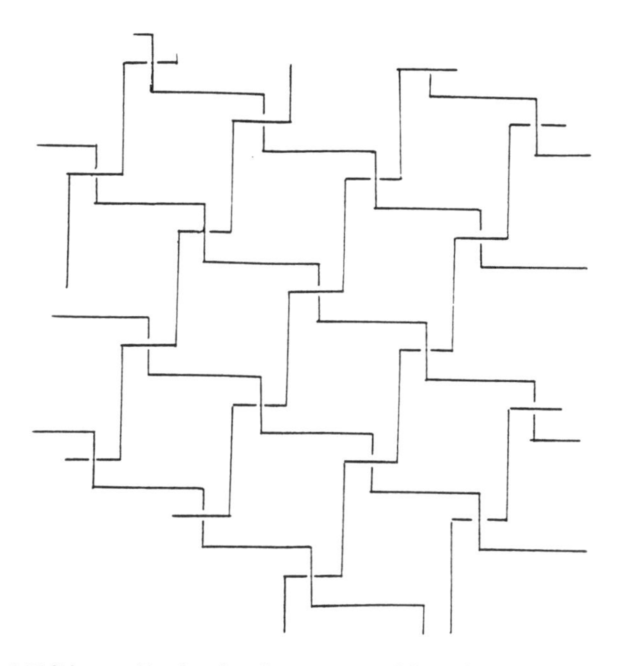

Figure 6-7 (b) ***Design having two- and four-fold rotocenters***

By comparison with Figure 6-7 (b) we note that the points where the lines interrupt, or pass over and under, each other are *two*-fold, not four-fold rotocenters. The points on the lines halfway between these two-fold rotocenters are not rotocenters at all in this example. In the spaces between the lines we still find four-fold rotocenters, but they are not all mutually equivalent in Figure 6-7 (b). In Figure 6-7 (a) such four-fold rotocenters were all equivalent to each other, being related by the two-fold rotocenters halfway between them. In Figure 6-7 (b) there are no rotocenters halfway between these four-fold rotocenters, and we do notice on careful inspection that the crossing patterns of the lines look different when viewed from the vantage points of two neighboring four-fold rotocenters. Clearly Figures 6-7 (a) and 6-7 (b) have the same symmetry, namely $244'$; one pattern is exactly as symmetrical as the other, even though, if Figures 6-7 (a) and 6-7 (b) are superimposed, the distance between rotocenters in 6-7 (a) will be seen to be smaller than in 6-7 (b).

NOTES

[1] Chorbachi, Wasma'a and Arthur L. Loeb: *Notation and Nomenclature* in *Symmetry of Structure*, Gy. Darvas and D. Nagy, eds. Budapest (1989).

VII

Enantiomorphy

In the previous chapter we noted that the mesh in the 236 system is a right triangle having angles of 30° and 60°. Adjacent meshes cannot be made to overlap each other completely by rotation in the plane; one needs to be flipped over out of the plane to be brought into coincidence with its neighbor. The operations which we have considered so far, rotation and translation, do not relate motifs which need to be flipped over to be brought into coincidence with each other. In Figure 7-1 we see two motifs which are mutually congruent but which are not related by rotation in the plane or by translation. They are, in point of fact, each others' mirror images; we call them *oppositely* congruent. Flipping one out of the plane to bring it into coincidence with its mirror image is called *reflection* ; an object and its mirror image are called a pair of *enantiomorphs* . (En-anti-o-morph in Greek means "in opposite form.")

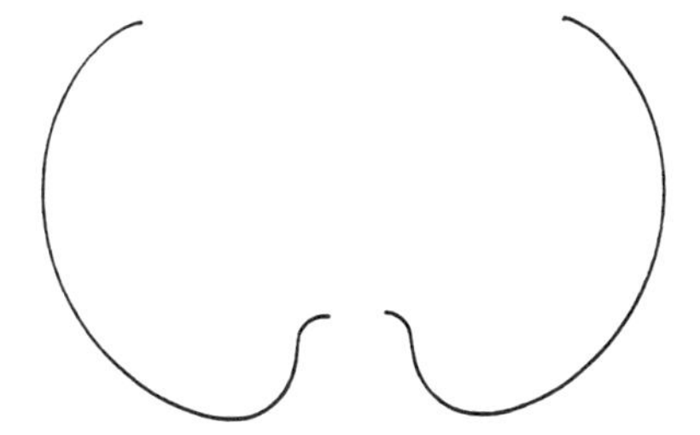

Figure 7-1 ***Enantiomorphy***

We are used to seeing our own image in the mirror. Our image in a mirror is not identical to our image on a photograph: They are a pair of enantiomorphs. When we look at our image in a mirror, we see that image directly opposite us. In Figure 7-2 we construct a two-dimensional analog: We see that it is not necessary to flip the motif out of the plane to obtain its enantiomorph. From every point in the motif a perpendicular is dropped on the "mirror line" and extended by the same length to generate

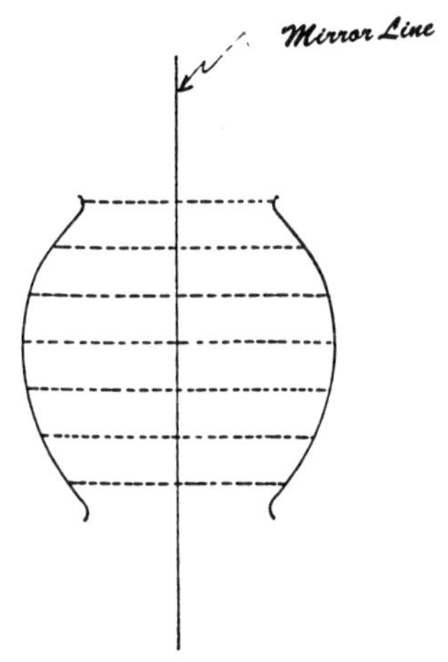

Figure 7-2 ***Construction of a mirror image***

its enantiomorph. The mirror line perpendicularly bisects the line joining every pair of enantiomorphic points.

In Figure 7-3 (a) we see a bowl and its mirror image: The two are virtually identical. However, in Figure 7-3 (b) we have combined the left hand side of the original bowl with its mirror image, and in Figure 7-3 (c) the right-hand side with its own mirror image. If the original bowl had been identical to its mirror image, then Figures 7-3 (b) and 7-3 (c) would have been identical. The perception of the difference between the original bowl and its mirror image is enhanced by the construction of Figures 7-3 (b) and 7-3 (c).

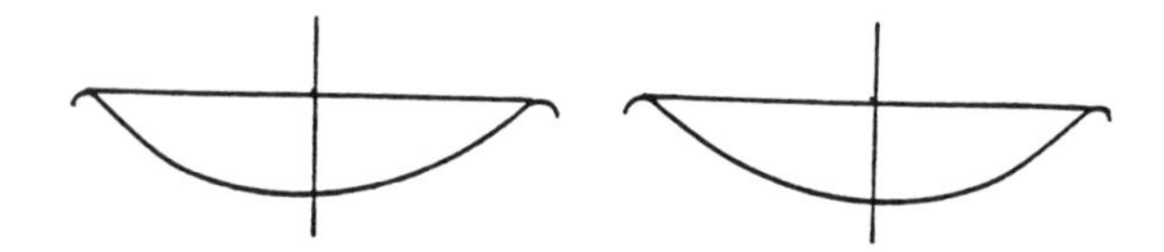

Figure 7-3 (a) ***A bowl and its mirror image***

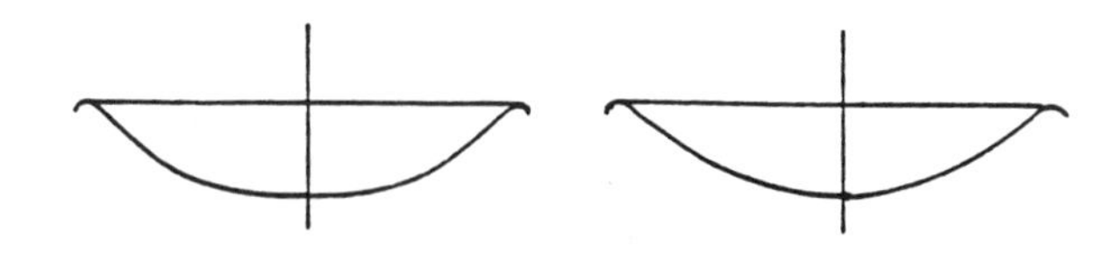

Figure 7-3 (b) and (c) ***The bowl shown in figure 7-3 (a) was not itself mirror-symmetrical***

The bowl in Figure 7-3 (a) looks almost symmetrical; we have shown through our construction just how it deviates from symmetry. Actual human faces are like Figure 7-3 (a): Although perceived as symmetrical, they rarely are. Moreover, our internal organs are asymmetrically located within a nearly symmetrical framework. By contrast, we note the sea urchin of Figure 7-4, which has rotational as well as mirror symmetry. Is this creature so much more "perfect" than the human organism? Why are humans actually asymmetrical? What caused the heart to move left of center in most humans? Does this asymmetry make humans less perfect? In point of fact, the problem with this question lies with the definition of "perfection." It turns out that only the simplest systems are completely symmetrical: Most complex systems are at most nearly symmetrical. In practical applications, we often make use of near-symmetry rather than perfect symmetry. Metals are ductile only because of imperfection in the alignment of their atoms. Semiconductors function as the result of impurities introduced into pure crystals.

We observed in Figures 7-3 (a), (b) and (c) that a pattern which is not quite symmetrical is perceived to be more symmetrical than it actually is. There appears to be a perceptive force toward symmetry: The tension created by this force makes nearly symmetrical patterns attractive.

Figure 7-4 *Sea Urchin (photography, David Caras)*

The reason for studying symmetry thus is not because perfect symmetry is something to be achieved, but because symmetrical patterns are easily perceived and recognized, hence can serve as a stabilizing frame of reference upon which perturbations may be superimposed to create tensions.

Although the *outlines* of adjacent meshes are enantiomorphic, their *contents* may or may not be related by mirror symmetry. If the contents of adjacent meshes are enantiomorphic, then the boundary separating these meshes will be a mirror line. The meshes for the systems 236, 244′ and 33′3″ were shown in Figures 6-3, 6-4 and 6-5. Figures 7-5 (a), (b) and (c) illustrate these respective symmetry combinations; adjacent meshes are not symmetrically related.

In the 244′ system adjacent meshes may be paired along their hypotenuse or along their right-angled side.

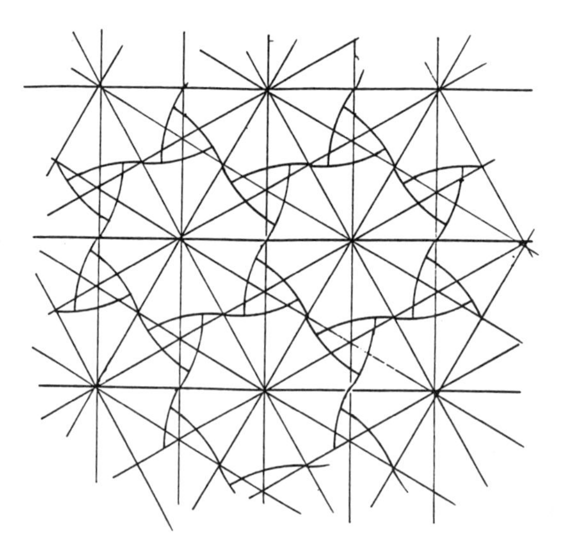

Figure 7-5 (a) ***236 Symmetry without enantiomorphy***

Q: *DRAW BOTH TYPES OF PAIRS OF MESHES, ALONG WITH THE INFORMATION CONTAINED IN THEM. FOR THE 236 SYSTEM THREE DISTINCT PAIRINGS ARE POSSIBLE. DRAW THESE TOGETHER WITH THE INFORMATION CONTENT OF EACH.*

By contrast, in Figures 7-6 (a), (b) and (c) adjacent meshes are enantiomorphically related; all rotocenters lie on mirrors. To distinguish the former three patterns from the latter, we underline the rotational symmetry value for all rotocenters lying on mirrors: $\underline{2}$, $\underline{3}$, $\underline{4}$, $\underline{6}$.

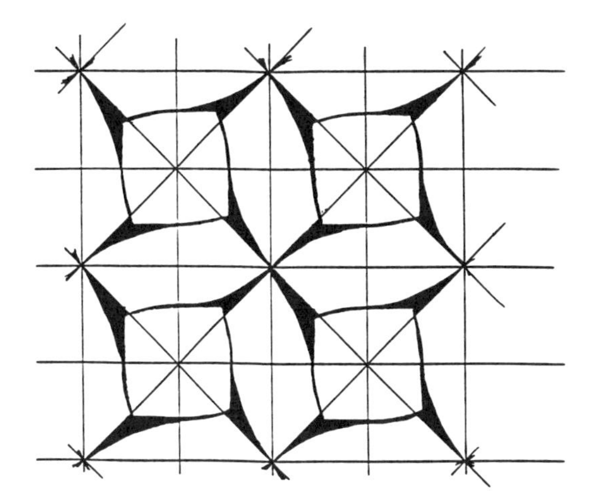

Figure 7-5 (b) ***244′ Symmetry without enantiomorphy***

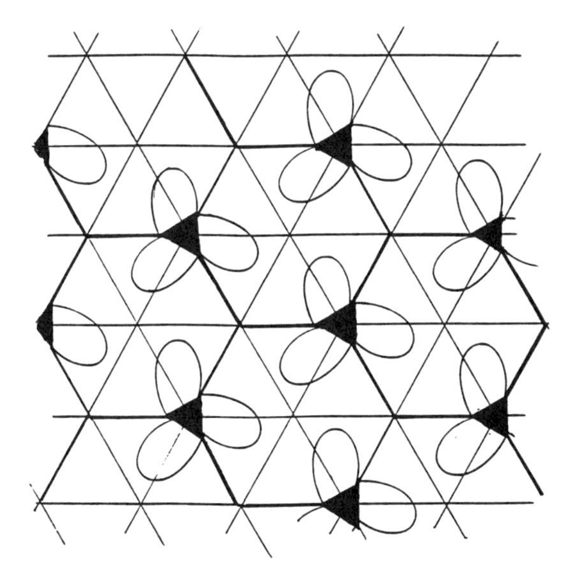

Figure 7-5 (c) ***33′ 3″ Symmetry without enantiomorphy***

Figure 7-6 (a) warrants some further consideration. It was generated from the mesh shown in Figure 7-7; an equilateral triangle having each vertex on one of the mesh lines. Even though this triangle has three-fold symmetry, its placement in the mesh is unsymmetrical, so that we may consider it as an unsymmetrical motif in this context. Reflection in the mesh

Figure 7-6 (a) 236 ***Symmetry having all rotocenters on mirrors***

Figure 7-6 (b) 244′ ***Symmetry having all rotocenters on mirrors***

boundaries and rotation around the rotocenters produced the 236 pattern of Figure 7-6 (a). The two-fold rotocenters are located in the rhombi, the three-fold rotocenters in the six-pointed stars and the six-fold rotocenters in the large hexagons. Note that the six-pointed stars contain only three-fold rotocenters in the context of the pattern, even though individually they are six-fold symmetrical.

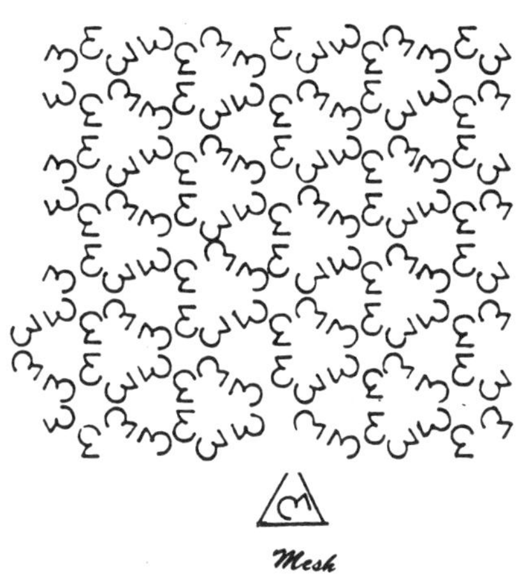

Figure 7-6 (c) ***"Triple Trinity": $\underline{33'3''}$ Symmetry having all rotocenters on mirrors***

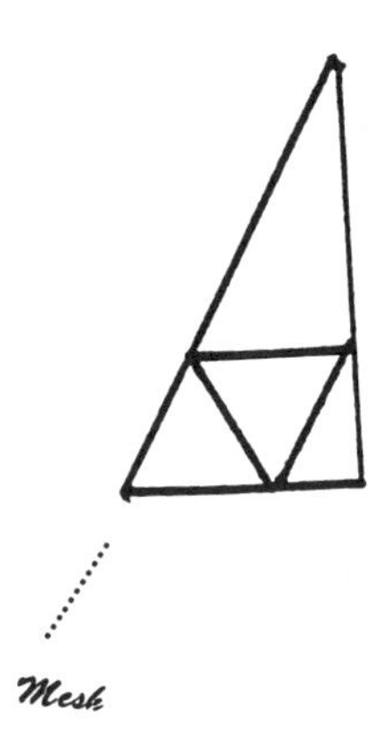

Figure 7-7 ***Motif and mesh for figure 7-6 (a)***

Closer inspection reveals that this pattern simply consists of superimposed large equilateral triangles whose edge length is six times that of the edge length of the original motif. Figure 7-6 (a) thus may be drawn as straight lines six motif-units long, interrupted by three units, then followed by further lines six units long. It is interesting that a pattern as complex as that in Figure 7-6 (a) should have been generated by a single triangle, and also that the resulting structure actually turns out to be hierarchical, being constituted of superimposed, interpenetrating large equilateral triangles.

In Figure 7-6 (c) ("Triple Trinity") the motif is the digit 3.

Q: *FIND THE THREE DISTINCT THREE-FOLD ROTOCENTERS.*

Let us consider the implication of a rotocenter lying on a mirror line. Figure 7-8 shows the Design Science logo. It was generated by placing an asymmetrical triangle on a mirror which also contained a three-fold rotocenter. Five additional triangles were accordingly generated, of which two are related to the first one by rotational symmetry, the others to these by mirror symmetry. We denote the symmetry of the Design Science logo as $\underline{3}$, the underline indicating that the rotocenter lies on a mirror line. This type of construction may be applied to any rotocenter on a mirror, regardless of its symmetry value. In general, a k-fold rotocenter on a mirror line is denoted $\underline{k}$.

Q: *PLACE A PIECE OF TRACING PAPER OVER THE DESIGN SCIENCE LOGO, AND DRAW THE LINES OF MIRROR REFLECTION (MIRROR LINES). HOW MANY MIRROR LINES DOES THE LOGO HAVE?*

The three-fold rotocenter on the mirror generated two additional mirrors, all three mirrors being at 120° to each other. The reader might wonder why we did not call the angle between the mirros 60° rather than 120°. Note, however, that, as one travels along one of the mirrors, it makes a difference whether one travels in one, or in the opposite, sense: Equivalent directions in any two of the three mirrors are oriented 120° apart.

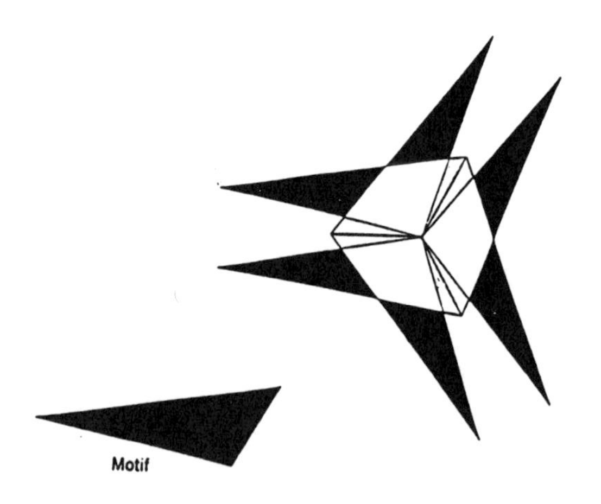

Figure 7-8 ***Design Science logo: A three-fold rotocenter on a mirror***

In Figure 7-9 (a) a two-fold rotocenter is located on a mirror line. A motif in the shape of a half-arrow is reflected in the mirror, and the original half-arrow as well as its mirror image are rotated 180° around the rotocenter. A new mirror is implied, passing through the rotocenter, perpendicular to the original mirror (Figure 7-9 (b)). This new mirror is distinct from the original one; each mirror, being bisected by the other, has itself mirror symmetry: Two opposite senses of travel along either mirror are equivalent.

We have encountered two different kinds of mirrors. Those for which two opposite senses of travel are distinct are called *polar*; those in which opposite directions are equivalent are *non-polar*. The findings for a two-fold and for a three-fold rotocenter on a mirror may be generalized as follows:

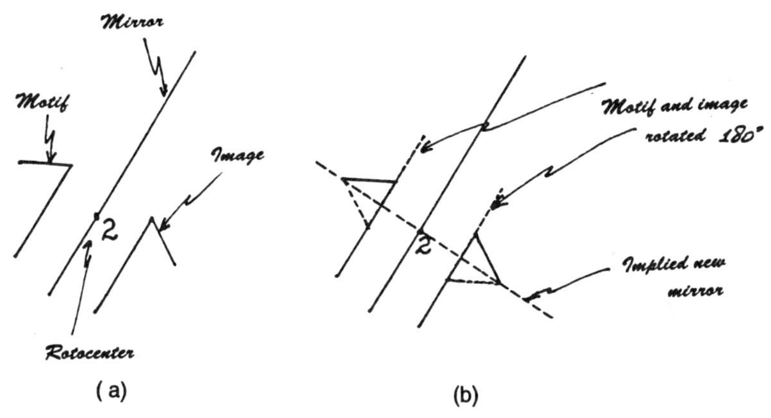

Figure 7-9 (a) and (b) ***A two-fold rotocenter on a mirror line***

Theorem 7-1: A k-fold rotocenter on a mirror implies the intersection of k mirrors at equal angles at the rotocenter. If k is odd, all mirrors are polar and mutually equivalent, but if k is even, the mirrors are non-polar and belong to two distinct sets.

When two mirrors M and M′ intersect at angle α (Figure 7-10), each will be reflected in the other. If we define an angular coordinate θ so that the mirror M is located at $\theta = 0$ and M′ at $\theta = \alpha$, then additional mirrors are implied as follows:

$$M \quad \text{at} \quad \theta = 2i\alpha, \quad \text{and} \quad M' \text{ at} \quad \theta = (2i+1)\alpha, \quad \text{where } i \text{ is an integer.}$$

Successive reflection of a motif in two adjacent mirrors produces a mirror image of a mirror image, which is directly congruent with the original. Therefore rotational symmetry is implied: The nearest pair of mutually congruent motifs are separated by an angle 2α. If α is chosen suitably, then M should reoccur exactly after 2π radians, i. e., when $2\pi = 2i\alpha$. Such a suitable choice would therefore be $\alpha = \pi/i$.

Q: *WHAT WOULD HAPPEN IF α WERE NOT CHOSEN SUITABLY?*

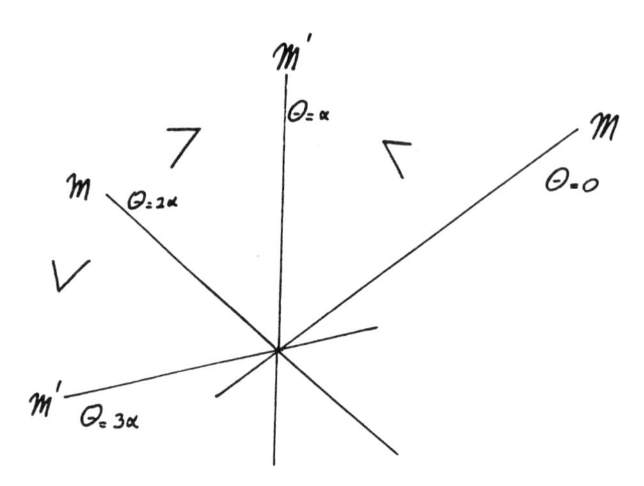

Figure 7-10 ***Intersecting mirrors***

When $\alpha = \pi/i$, mutually congruent motifs occur at intervals $2\alpha = 2\pi/i$, so that i-fold symmetry is implied. It is also interesting to consider what would happen if a mirror M′ were implied at $\theta = 2\pi$. This would be the case, for instance, if α were chosen to be $2\pi/3$; an example would again be the Design Science logo (Figure 7-8). M and M′ would then be equivalent, so that, as we have seen, there is only one type of mirror, which is polar. A mirror M′ is generally implied at $\theta = 2\pi$ if α is chosen such that $2\pi = (2j+1)\alpha$, i. e., $\alpha = 2\pi/(2j+1)$, j being an integer. For instance, for the Design Science logo there are equivalent mirrors making angles of $2\pi/3$ radians with each other and the implied rotational symmetry is three-fold. If α had been chosen such that $j = 2$, then $\alpha = 2\pi/5$; there would then be five polar mirrors and five-fold implied rotational symmetry. These results can be summarized as follows:

Theorem 7-2: Two distinct mirrors intersecting at an angle π/k, where k is an integer, imply a center of k-fold rotational symmetry at their intersection; there will be k mirrors, which will be non-polar and of two distinct types if k is even and polar and mutually equivalent if k is odd.

Two mirrors intersecting at an angle $2\pi/(2k+1)$ imply a center of $(2k+1)$-fold rotational symmetry at their intersection; there will be $(2k+1)$ mutually equivalent polar mirrors.

We saw in the previous chapter that when more than a single rotocenter coexist in a plane, their symmetry values are limited to the values 2, 3, 4, 6 and ∞. Therefore mirror lines intersecting at such rotocenters will intersect at angles 30°, 45°, 60°, 90° and 120°, all of which are special cases of the expressions π/k or $2\pi/(2k+1)$.

For completeness, we should also consider what is implied when a single pair of mirror lines intersect at an angle that cannot be expressed in that form. For instance, take two mirrors M and M′ intersecting at an angle of 65°. There will then be mirrors equivalent to M at angular positions $\theta = 130°, 260°$ and 390° (which is the same as 30°). Mirrors equivalent to M′ occur at $\theta = 65°, 195°, 325°$ and 455° (which is the same as 95°). Mirrors M at $\theta = 0°$ and M′ at $\theta = 30°$ together imply M at $\theta = 60°$; in turn M at $\theta = 60°$ and M′ at 65° together imply mirrors at intervals of 5°, i. e., $\alpha = \pi/36$ radians, to which Theorems 7-1 and 7-2 will apply. Accordingly, two mirror lines intersecting at 65° imply 36-fold rotational symmetry! Similarly, any value of α which is a *rational* fraction of 2π radians will eventually imply an angle to which these theorems apply. If α is an *irrational* fraction of 2π radians, then eventually mirror lines will be implied whose angular separation is arbitrarily small; if we extend the postulate of closest approach to mirrors, postulating for every pattern having mirror symmetry a smallest angle between mirror lines, then an irrational fraction of 2π radians is excluded.

Rotocenters may themselves be enantiomorphically paired. In Figure 7-11 (a) we show a pair of two-fold rotocenters on opposite sides of a mirror line; we label these 2 and $\tilde{2}$ respectively.

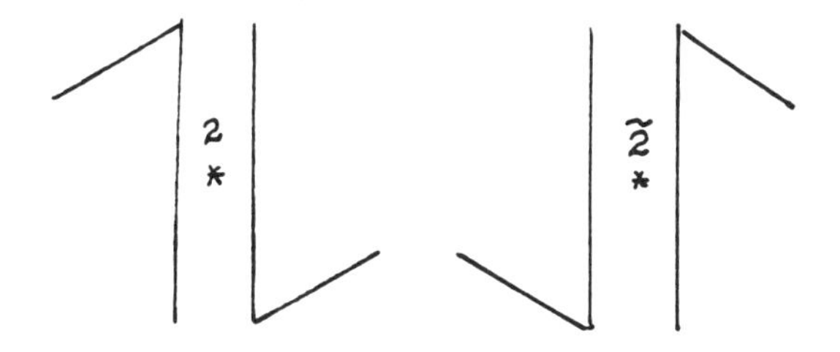

Figure 7-11 (a) ***A pair of enantiomorphic two-fold rotocenters***

Q: *WHAT DOES THE COEXISTENCE OF 2 AND $\tilde{2}$ IMPLY? IS THERE JUST ONE MIRROR, OR ARE ADDITIONAL MIRRORS IMPLIED?*

Just as in the case of two independent two-fold rotocenters 2 and 2′, this pair of enantiomorphically paired rotocenters generates a row of alternating rotocenters, implying translational symmetry; we denote this special case of the $22'\infty$ system $2\tilde{2}\infty$ (Figure 7-11 (b)). There is a mirror line halfway between each pair of adjacent two-fold rotocenters.

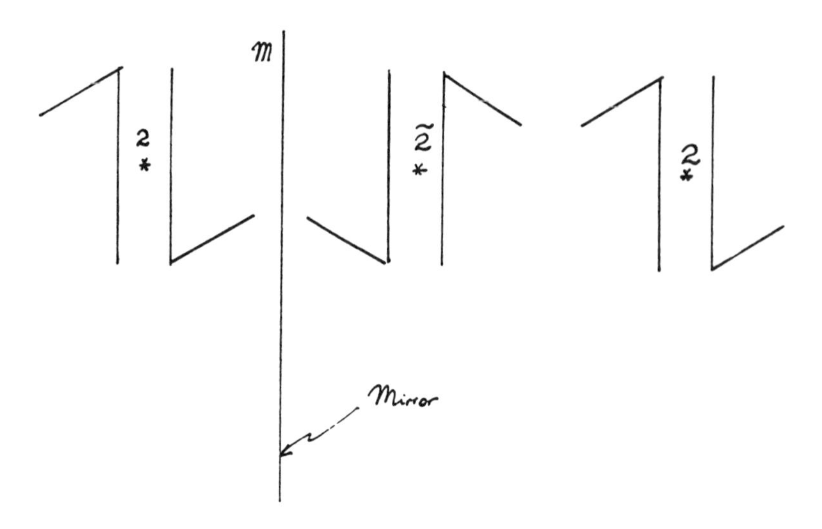

Figure 7-11 (b) ***Additional symmetry elements generated by the pair of rotocenters in figure 7-11 (a)***

We should refine our nomenclature at this point. Two rotocenters whose environments are identical (even if in different orientations) belong to the same *rotocomplex*, and are said to be *congruent* if they are related by rotational or translational symmetry, and *enantiomorphs* if related by reflection symmetry. Rotocenters which are either congruent or enantiomorphic are called equivalent; if not equivalent, they are called *distinct*. Schematically, we can summarize these distinctions as follows:

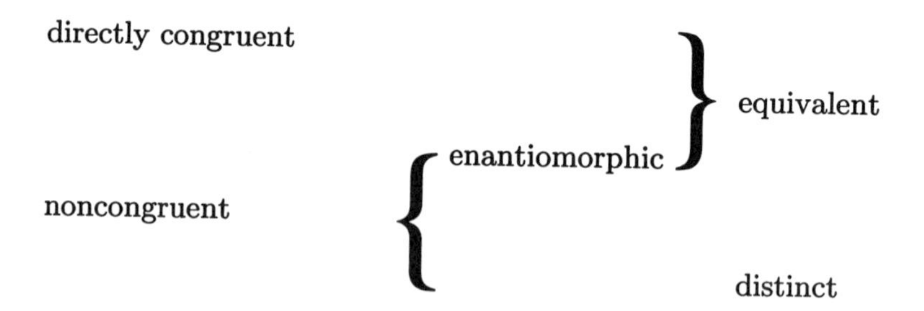

Distinct rotocenters having the same symmetry value are distinguished from one another by primes (e. g., 3, $3'$ and $3''$; enantiomorphs are distinguished from each other by a tilde (e. g., 2 and $\tilde{2}$). The derivation of the diophantine equation was based on the fact that all A_ks were mutually congruent, as were all B_ls. Accordingly, all noncongruent rotocenters, whether enantiomorphs or distinct, take up two places in the diophantine equation: Two enantiomorphic four-fold rotocenters would have $l = 4$ as well as $m = 4$.

Q: *WHAT DOES THE COEXISTENCE OF A ROTOCENTER AND A MIRROR LINE IMPLY?*

Since a k-fold rotocenter located off a mirror line generates as its image another rotocenter, $\tilde{k}, k$ and $\tilde{k}$ must together satisfy the diophantine equation, so that k is limited to the systems having distinct rotocenters of the same symmetry value: $1\infty\infty', 22'\infty, 244'$, and $33'3''$; the system 236 does not permit rotocenters which do not lie on the mirror lines, as neither two, three or six-fold rotocenters in this system are permitted to have distinct images $\tilde{2}, \tilde{3}$ or $\tilde{6}$ to coexist with. Accordingly, enantiomorphy in the 236 system is permitted only if all rotocenters lie on mirror lines (the mesh boundaries). Thus, in the 236 system we have only two possibilities: 236 (no enantiomorphy), 7-5 (a) and $\underline{236}$ (all rotocenters on mirror lines, Figure 7-6 (a)). Thanks to the diophantine equation, we shall only need to consider the interaction of a mirror line with two, three or four-fold rotocenters not on the mirror line.

As we saw above, a two-fold rotocenter located a finite distance from a mirror line implies its enantiomorph at an equal distance on the opposite side of the mirror line. The pair of enantiomorphic two-fold rotocenters 2 and $\tilde{2}$ implies a ribbon or frieze pattern, denoted by $2\tilde{2}\infty$ (Figure 7-11 (b)).

Mirror symmetry is not the only way in which enantiomorphs may be related. Figure 7-12 (a) shows the trail which might have been left by a snow-shoe walker in the snow. The prints of left and right shoes are certainly enantiomorphs, but there is no mirror line relating them, because they are not directly opposite each other. In Figure 7-12 (b) we have drawn a set of line segments joining the heels of successive footprints. Next we joined the centers of these line segments by a straight line. This line is called a *glide line* : A pattern which remains unchanged when mirror reflected in a glide line and then translated parallel to that line is said to have *glide symmetry*. The distance through which the pattern has to be translated after the mirror reflection to be brought into coincidence with its original is called the *glide component*. Mirror symmetry may be considered a special case of glide symmetry having zero glide component.

Q: *IN FIGURE 7-13 WE SHOW A COLLECTION OF PATTERNS, SOME OF WHICH HAVE GLIDE SYMMETRY, WHILE OTHERS DO NOT. THE READER IS INVITED TO DETERMINE WHICH DO, AND WHICH DO NOT, HAVE GLIDE SYMMETRY, AND WHY. (For solutions, cf. Appendix I).*

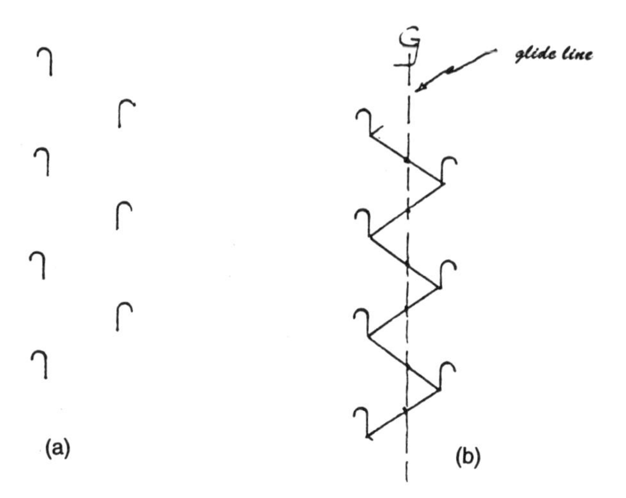

Figure 7-12 (a) and (b) ***Glide symmetry***

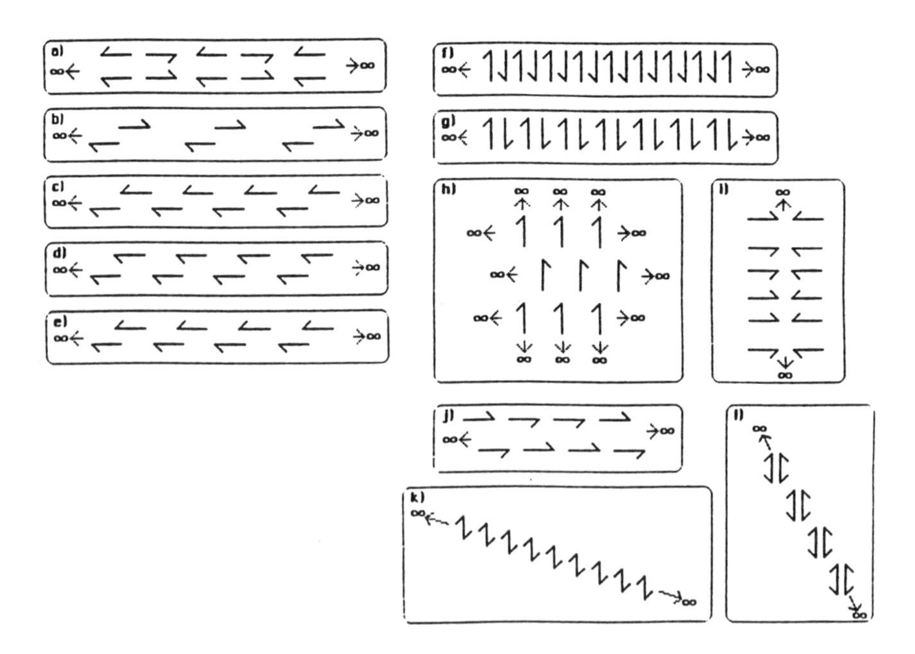

Figure 7-13 ***Which of these patterns have glide symmetry?***

Let us consider the interaction between a k-fold rotocenter and a glide line. The only rotocenters which can coexist with enantiomorphs to which they are not directly congruent (in other words, which do not lie on mirror lines) are two, three and four-fold ones. In Figure 7-14 (a) we have drawn a two-fold rotocenter together with a glide line, and the image of the two-fold rotocenter in the glide line.

Q: *DO YOU OBSERVE ANY ADDITIONAL MIRROR OR GLIDE LINES? IF SO, HOW MANY? ARE THESE TWO THE ONLY TWO-FOLD ROTOCENTERS? OR ARE ADDITIONAL ONES IMPLIED? HOW MANY?*

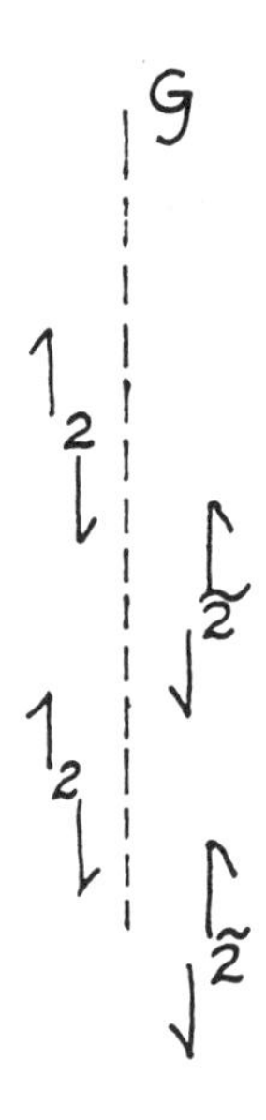

Figure 7-14 (a) ***Two-fold rotocenter near a glide reflection line***

It is observed that at the point equidistant from and collinear with the two rotocenters the glide line is perpendicularly intersected by a second glide line (Figure 7-14 (b)). The two glide lines must be glide-reflected into each other while the two-fold rotocenters are glide-reflected into them as well to generate the grid pattern of Figure 7-14 (c).

Recalling that a row of mutually congruent two-fold rotocenters located on a common straight line imply two-fold rotocenters halfway between them, we place additional rotocenters in the grid formed by the glide lines. These additional centers are distinct from the ones already entered in the grid, and also are enantiomorphically paired. The resulting configuration is accordingly labeled $2\tilde{2}2'\tilde{2}$.

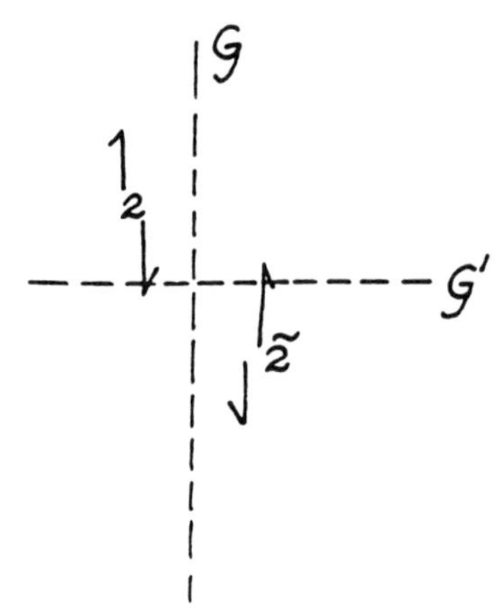

Figure 7-14 (b) ***The configuration of figure 7-14 (a) implies a second glide line***

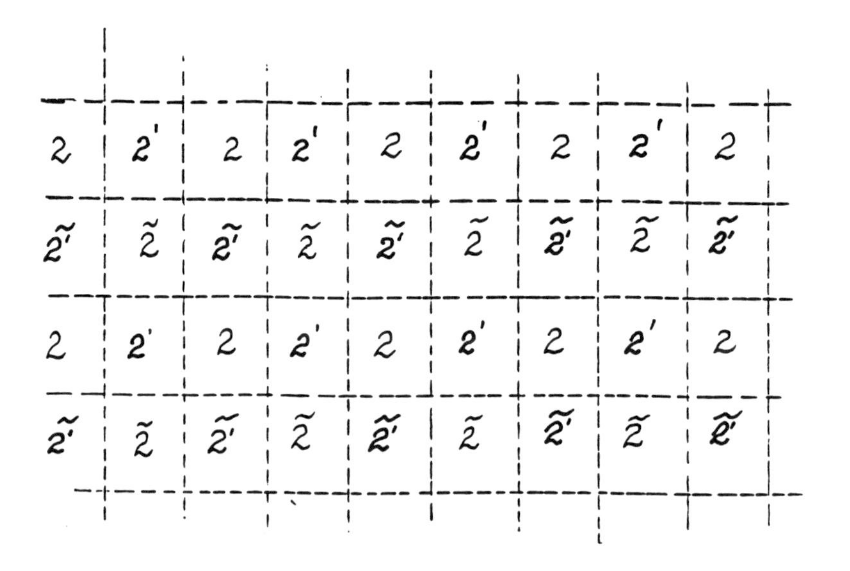

Figure 7-14 (c) ***The configuration*** $2\tilde{2}2'\tilde{2}$

There is also the possibility that the second reflection line implied by the coexistence of a two-fold rotocenter with a glide line is a mirror line (Figure 7-15 (a)). In that case, this mirror line will be reflected in the glide line to generate a set of mutually parallel mirror lines perpendicular to the glide line, and the orginal rotocenter generates collinear mutually enantiomorphic two-fold rotocenters labeled $2\tilde{2}\infty$, (Figure 7-15 (b)). This symmetry configuration is illustrated by the design of Figure 7-15 (c).

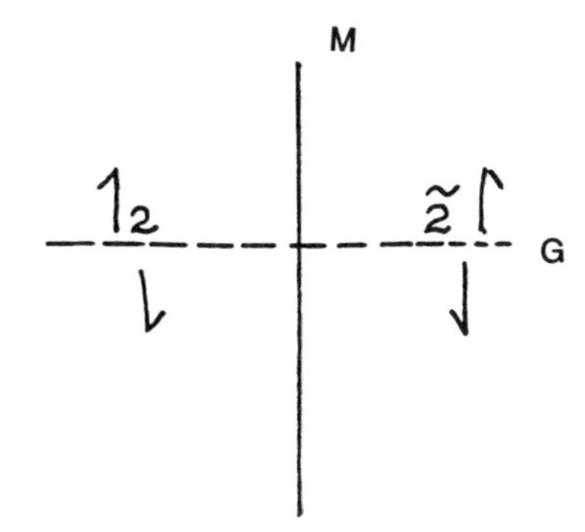

Figure 7-15 (a) *Implied mirror line*

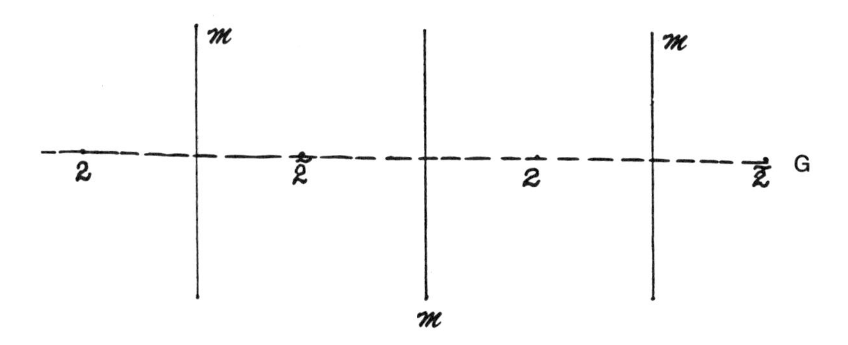

Figure 7-15 (b) $2\tilde{2}\infty$ *configuration*

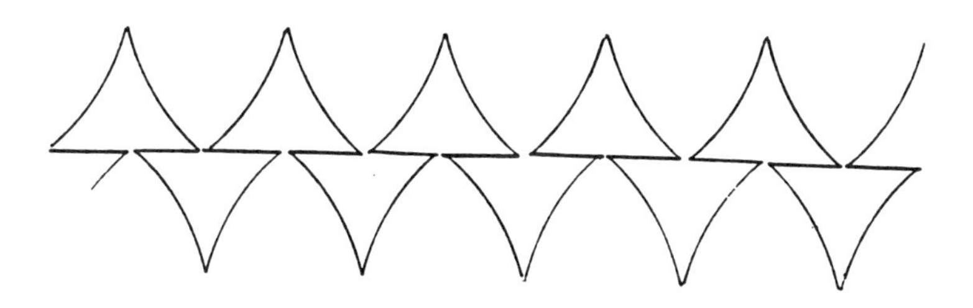

Figure 7-15 (c) *Pattern having symmetry* $2\tilde{2}\infty$

We have thus discovered

Theorem 7-3: The coexistence of a two-fold rotocenter and a mirror a finite distance apart implies an infinitely extended row of alternating mutually enantiomorphic rotocenters joined by a glide-reflection line. The glide line perpendicularly intersects mirror lines halfway between adjacent enantiomorphically paired rotocenters. The configuration so generated is labeled $2\tilde{2}\infty$.

A two-fold rotocenter on a glide line implies the same configuration $2\tilde{2}\infty$.

Since a two-fold rotocenter *near* but not *on* a mirror line generates a row of two-fold rotocenters joined by a glide line, and since a two-fold rotocenter on a glide line generates the same configuration, it would be tempting to conclude that a pair of enantiomorphically paired two-fold rotocenters implies that same configuration. Figure 7-14 (c) foils that temptation, for this figure was generated by a pair of enantiomorphically paired two-fold rotocenters! Therefore:

Theorem 7-4: The coexistence of an enantiomorphic pair of two-fold rotocenters implies either the monoperiodic configuration $2\tilde{2}\infty$ or the diperiodic configuration $22'2\tilde{2}'$, in which glide lines are mutually parallel or perpendicular.

A three-fold rotocenter off a mirror line implies its enantiomorph on the opposite side of the mirror line. According to the diophantine equation the pair $3\tilde{3}$ implies a third three-fold rotocenter, which cannot have an enantiomorphic partner and therefore must itself lie on a mirror line. Hence:

Theorem 7-5: The existence of a three-fold rotocenter off a mirror line implies the configuration $3\tilde{3}\underline{3}'$.

Theorem 7-1 tells us that three mutually equivalent mirror lines will intersect at the $\underline{3}'$ rotocenters; these mirror lines pass halfway between the 3 and $\tilde{3}$ rotocenters. This configuration is illustrated in Figure 7-16: At the top a curved motif is visible together with its enantiomorph, each growing out of one of a pair of enantiomorphic three-fold rotocenters. The entire pattern is defined by these elements and is generated as a result of their interactions.

Q: *FIND THE MIRROR AND GLIDE LINES IN FIGURE 7-16.*

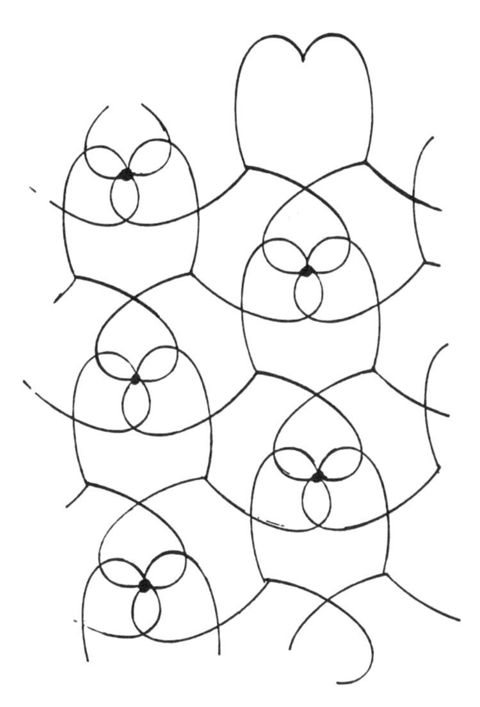

Figure 7-16 ***Pattern having the symmetry configuration*** $3\tilde{3}\underline{3}'$.

A four-fold rotocenter off a mirror line analogously implies its enantiomorph, and this enantiomorphic pair in turn implies a two-fold rotocenter, which necessarily lies on a mirror line.

Theorem 7-6: The existence of a mirror line off a four-fold rotocenter implies the configuration $\underline{2}4\tilde{4}$.

According to Theorem 7-1, two mirror lines intersect perpendicularly at the two-fold rotocenters, as illustrated in Figure 7-17: The "bow-ties" lie across the two-fold rotocenters, whereas the squares are centered on the mutually enantiomorphic four-fold rotocenters. Observe that kite-like quadrilaterals (half bowties) are attached to the corners of the squares, and that these kites will rotate clockwise or counter-clockwise around the enantiomorphic rotocenters. The lower portion of the illustration shows the frame of reference for this figure: The generating motif here was a small triangle, one quarter of a bow-tie.

Q: *FIND ALL MIRROR AND GLIDE LINES IN FIGURE 7-17.*

Figure 7-17 ***Patterns having the symmetry configuration*** $\underline{2}4\tilde{4}$.

Q: *USING THE FRAMEWORK OF FIGURE 7-17. CREATE OTHER INTERESTING PATTERNS BY COLORING DIFFERENT REGIONS ENCLOSED BY THE BLACK LINES.*

VIII

Symmetry Elements in the Plane

Recall from Chapters V and VI that the following combinations of rotocenters may coexist in the plane:

1. a single rotocenter
2. three non-congruent rotocomplexes:
   $1\infty\infty'$
   $22'\infty$
   236
   $244'$
   $33'3''$
3. four non-congruent rotocomplexes:
   $22'2''2'''$

Q: *DRAW THE CONFIGURATION OF ROTOCENTERS AND MESHES FOR EACH OF THESE.*

Each of these systems will now be further subdivided according to their enantiomorphy. Using the results of our experiments of Chapter VII, we shall examine each of the combinations listed above.

1. A single rotocenter. As we saw, a single rotocenter may lie on a mirror regardless of its symmetry value: As long as the symmetry value, denoted by k, is finite, there are only two possibilities, k and $\underline{k}$. For completeness we should consider $k = 1$, which has no symmetry at all, and $\underline{1}$, a single mirror. The Design Science logo (Figure 7-8) is an example of $\underline{k}$.

In addition, k may be infinite, in which case there is translational symmetry in a single direction; in the absence of enantiomorphy the symmetry is designated ∞ (Figure 8-1).

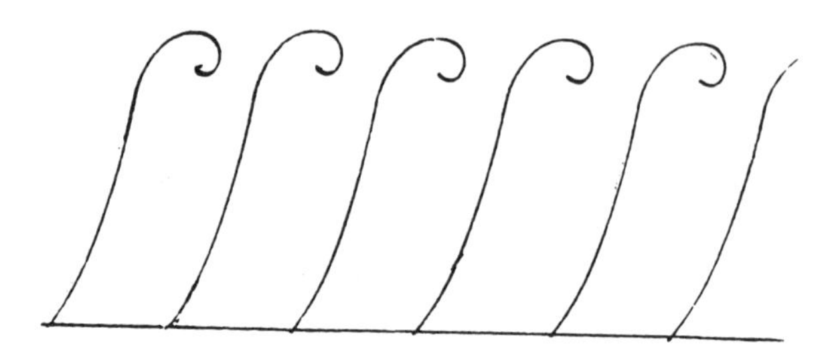

Figure 8-1 ***Translational symmetry in a single direction:*** ∞

Q: *CONSIDER THE FOLLOWING PATTERNS: a) A ROW OF IDENTICAL TREES SEEN ACROSS A RIVER; b) A MAN IN A BARBER'S CHAIR FACING A MIRROR WHILE ON THE OPPOSITE WALL, IN BACK OF HIM THERE IS ALSO A MIRROR PARALLEL TO THE FIRST MIRROR; AND c) SNOWSHOE TRACKS IN THE SNOW. WHAT CONFIGURATION OF SYMMETRY ELEMENTS WOULD BEST DESCRIBE EACH OF THESE PATTERNS? DO ALL OF THEM CORRESPOND TO INFINITE-FOLD ROTATIONAL SYMMETRY?*

There may be a single mirror along the direction of translational symmetry (Figure 8-2) or a glide line along the direction of translational symmetry (the "snowshoe" pattern); these two configurations are denoted ∞m and ∞g respectively (Figure 7-12 (a) represents the latter).

Finally, there may be a mirror perpendicular to the direction of translational symmetry; the translational symmetry itself will generate infinitely many mirror lines parallel to this first one. As shown in Figure 8-3, halfway between these mirrors there is a set of mirrors distinct from the first set.

Q: *CAN YOU FIGURE OUT THE MOTIF WHICH GENERATED THIS PATTERN BY SUCCESSIVE REFLECTIONS? DID IT PASS "THROUGHT THE LOOKING GLASS?"*

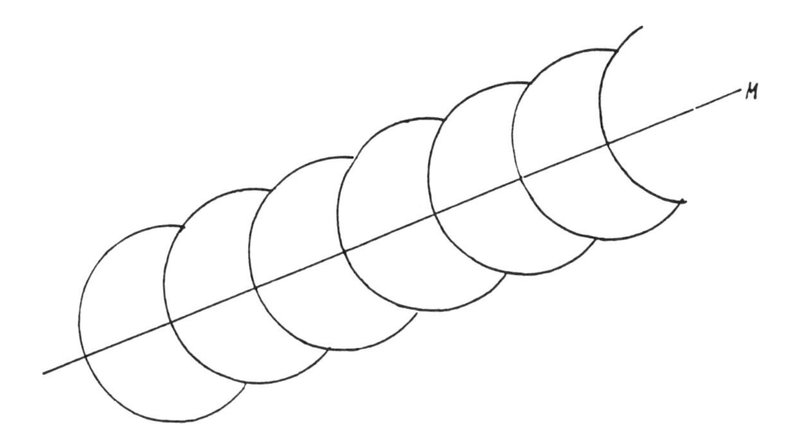

Figure 8-2 ***Mirror line parallel to the direction of translational symmetry:*** ∞m

This configuration is exemplified by the pattern generated at a barber's or a hairdresser's where there are two parallel mirrors on opposite walls. The customer will see an infinitely extended row of images alternately facing mirrors or turned away from them. This configuration is labeled $\infty mm'$. Any reflection line other than parallel to or perpendicular to the direction of translational symmetry would generate translational symmetry in more than a single direction and therefore would not belong to the system ∞. Thus the symmetry configurations having at most a single rotocenter are

$$1,\ m,\ k,\ \underline{k},\ \infty,\ \infty mm',\ \infty m \quad \text{and} \quad \infty g\,.$$

2. *Three non-congruent rotocomplexes.* The first combination to consider is $1\infty\infty'$. Figure 3-5 illustrates this symmetry: Translation in two independent directions, no enantiomorphy. Since we noted that reflection lines intersecting at any finite angle will imply a finite rotational symmetry, this system will permit only mutually parallel reflection lines.

Q: *DRAW ALL COMBINATIONS OF TWO PARALLEL REFLECTION (MIRROR OR GLIDE) LINES AND A STRING OF MOTIFS SYMBOLIZING TRANSLATIONAL SYMMETRY ALONG ONE OF THE REFLECTION LINES. COMPLETE THE PATTERN, DRAWING EVERY MOTIF AND SYMMETRY ELEMENT IMPLIED.*

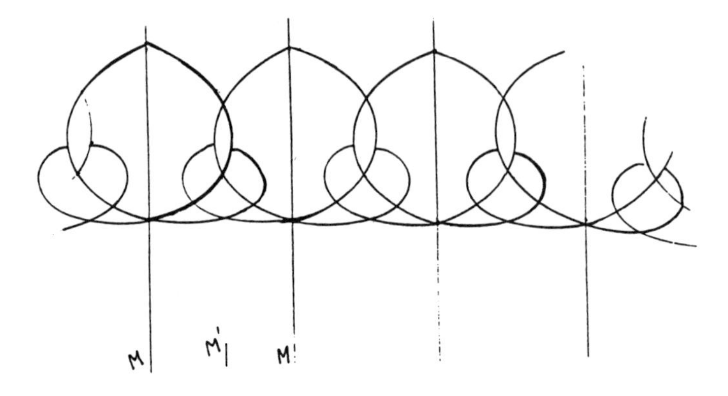

Figure 8-3 ***Mirrors perpendicular to the direction of translational symmetry:*** $\infty mm'$

There may be two parallel mirrors, $1\infty\infty' mm'$ (Figure 8-4), a mirror line parallel to a glide line $1\infty\infty' mg$ (Figure 8-5), and two parallel glide lines $1\infty\infty' gg'$ (Figure 8-6).

Next is the system $22'\infty$. Figure 8-7 shows the two-fold rotational symmetry without enantiomorphy. Enantiomorphy may occur when all rotocenters lie on mirrors, or 2 and $2'$ may be enantiomorphically paired. In the former case a mirror line would join all the rotocenters; Theorem 7-1 tells us that in that case there will also be a mirror perpendicularly intersecting that first mirror at every rotocenter. This configuration is labeled $\underline{22'\infty}$ (Figure 8-8). When 2 and $2'$ are enantiomorphically paired, Theorem 7-4 tells us that the only *monoperiodic* configuration will be $2\tilde{2}\infty$, a glide line joining the enantiomorphically paired rotocenters, of which every segment joining nearest rotocenters is perpendicularly bisected by a mirror (Figure 8-9). Note that these mirrors are polar and each mirror is oriented in a sense opposite to that of its two nearest neighbors. The other configuration implied by the coexistence of 2 and $\tilde{2}$ according to Theorem 7-4 is diperiodic, and therefore will be considered later.

In Chapter VII we already showed that in the 236 system (for 236 without enantiomorphy, cf. Figure 7-5 (a)) enantiomorphy is possible only

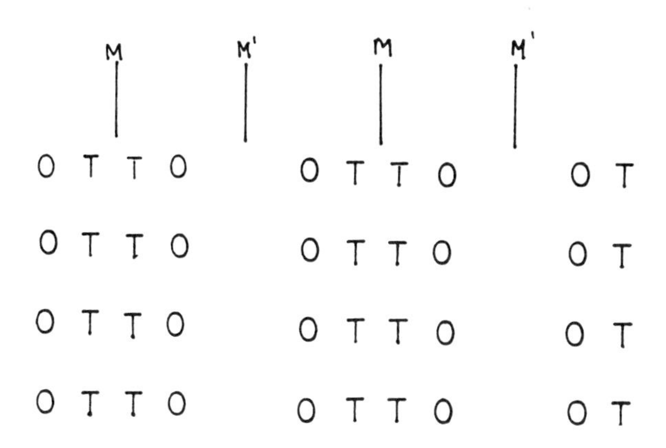

Figure 8-4 ***A pattern having the configuration*** $1\infty\infty' mm'$

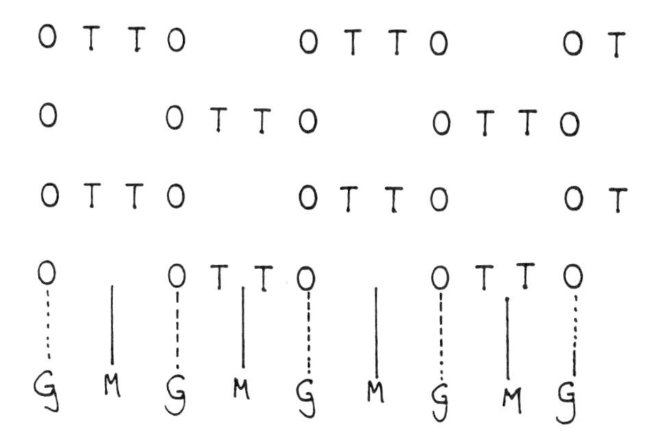

Figure 8-5 ***A pattern having the configuration*** $1\infty\infty' mg$

if all rotocenters lie on mirrors: $\underline{236}$ (cf. Figure 7-6 (a)). In the $244'$ system, however, either all rotocenters lie on mirrors ($\underline{244'}$, Figure 7-6 (b)), or the two non-congruent four-fold rotocomplexes may be enantiomorphically paired. In the latter case, since there is only a single two-fold rotocomplex, all two-fold rotocenters must lie on mirrors ($\underline{2}4\tilde{4}$, Figure 8-10). Mirrors join

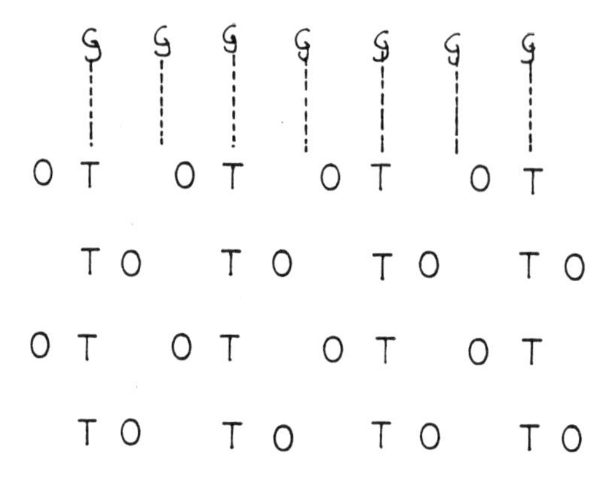

Figure 8-6 ***The configuration*** $1\infty\infty' gg'$

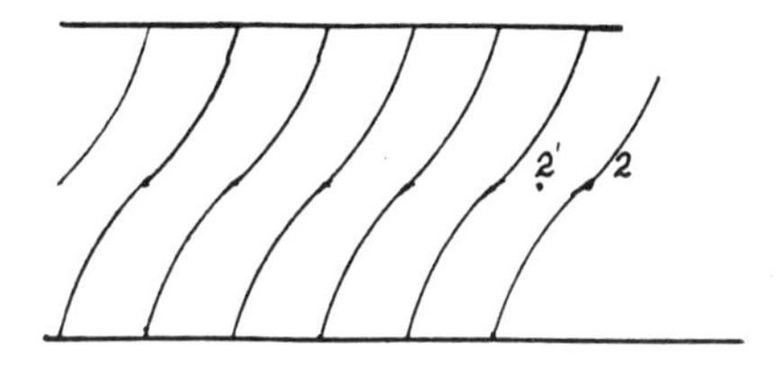

Figure 8-7 $22'\infty$ ***: No enantiomorphy***

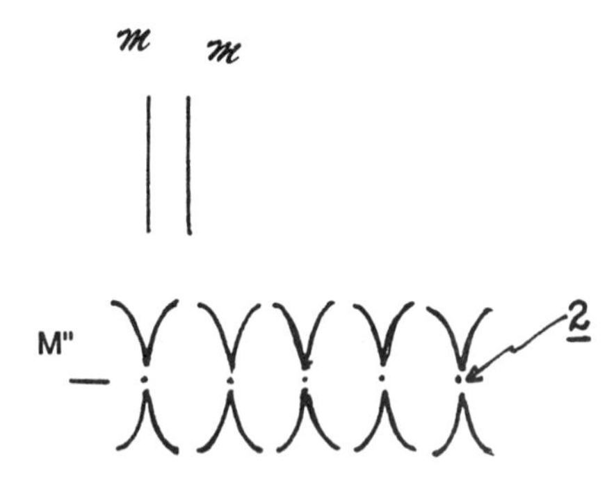

Figure 8-8 ***The configuration*** $\underline{22'\infty}$

Figure 8-9 ***The configuration*** $2\tilde{2}\infty$

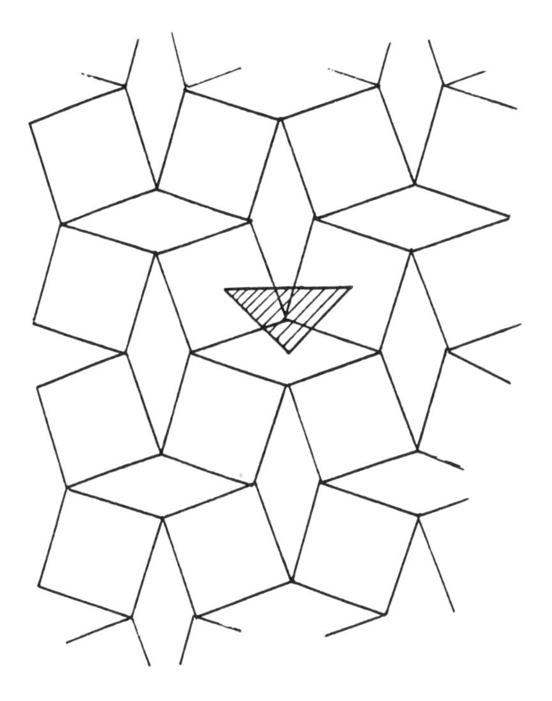

Figure 8-10 ***A pattern having the configuration*** $\underline{2}4\tilde{4}$

the two-fold rotocenters located at the centers of the rhombuses, passing <u>*between*</u> the four-fold rotocenters, hence bisecting the meshes at the centers of the square. (One mesh is shown cross hatched in Figure 8-10.) In the configuration $\underline{2}4\tilde{4}$, therefore, the meshes are each divided in half by mirror lines, one half being the mirror image of the other, while the contents of adjacent meshes are quite distinct.

Similarly, in the system $33'3''$ (cf. Figure 7-5 (c)), all rotocenters may lie on mirrors ($\underline{33'3''}$, Figure 7-6 (c)), or two rotocomplexes may be enantiomorphically paired while the third lies on mirrors ($\underline{3}'3\tilde{3}$, Figure 7-16). Here the mesh is an equilateral triangle bisected by a mirror.

Q: *TRACE THE MESH AND MIRROR ON A PIECE OF TRANSPARENT PAPER.*

Summarizing all configurations having three rotocomplexes:

$1\infty\infty'$, $1\infty\infty'mm'$, $1\infty\infty'mg$, $1\infty\infty'gg'$

$22'\infty$, $\underline{22'\infty}$, $\underline{2\tilde{2}\infty}$

236, $\underline{236}$

$244'$, $\underline{244'}$, $\underline{2}4\tilde{4}$

$33'3''$, $\underline{33'3''}$, $\underline{3'3\tilde{3}}$

3. Four non-congruent rotocomplexes. Here all rotocenters are two-fold and belong to four distinct rotocomplexes: $22'2''2'''$ (cf. Figure 1-3). All rotocenters may lie on mirror lines, which, according to Theorem 7-1, intersect perpendicularly at the rotocenters ($\underline{22'2''2'''}$, Figure 8-11). Another possibility is that two of the rotocomplexes are enantiomorphically paired, whereas the other two lie on mirror lines. Theorem 7-1 tells us that the rotocomplexes located on mirror lines imply a grid of mutually parallel and perpendicular mirror lines, whereas according to Theorem 7-4 the enantiomorphically paired two-fold rotocomplexes imply a grid of mutually perpendicular glide lines. As a result, there are alternating mirror and glide lines in two mutually perpendicular directions ($2\tilde{2}\underline{2'2''}$, Figure 8-12).

Finally, all four rotocomplexes may be enantiomorphically paired: $2\tilde{2}2'\tilde{2}'$. From Theorem 7-4 we learn that there are in this case two possibilities. The combination $2\tilde{2}$, as we saw in the instance of $22'\infty$, implies a row of alternating 2 and $\tilde{2}$ joined by a glide line, with mirrors perpendicularly bisecting the line segments joining adjacent rotocenters. Thus $2\tilde{2}2'\tilde{2}'$ may be considered a combination of parallel strings of two-fold rotocenters with the mirrors lined up (Figure 8-13 (a) and (b)). This configuration is designated $2\tilde{2}2'\tilde{2}'m/g$, the slash (/) indicating that all mirrors are perpendicular to all glide lines.

The second potential configuration implied by Theorem 7-4 is a grid of mutually parallel and perpendicular glide lines (Figure 8-14), labeled $2\tilde{2}2'\tilde{2}'g/g'$. There are therefore five configurations having four distinct rotocomplexes:

$$22'2''2''',\ \underline{22'2''2'''},\ 2\tilde{2}\underline{2'2''},\ 2\tilde{2}2'\tilde{2}'m/g, \quad \text{and} \quad 2\tilde{2}2'\tilde{2}'g/g'.$$

Figure 8-11 ***A pattern having the configuration*** $\underline{22'2''2'''}$

Q: *PLACE TRANSPARENT PAPER OVER FIGURES 8-12, 8-13, 8-14 AND 8-15, AND TRACE ALL REFLECTION LINES, MESHES AND ROTOCENTERS.*

All symmetry configurations are summarized in Table 8-1. Of these, seven ($\infty, \infty mm', \infty m, \infty g, 22'\infty, \underline{22'\infty}$ and $\tilde{2}2\infty$) are monoperiodic, four ($1, 1m, k, \underline{k}$) have at most a single finite rotocenter, and the remaining seventeen are diperiodic. Of all diperiodic configurations having finite rotational symmetry and enantiomorphy, $2\tilde{2}2'\tilde{2}'g/g'$ is the only one in which there are glide lines but no mirror lines.

Q: *WHICH CONFIGURATIONS HAVE ALL ROTOCENTERS ON MIRRORS? WHICH HAVE NO ENANTIOMORPHY? WHICH HAVE SOME ROTOCENTERS ON MIRRORS, OTHERS ENANTIOMORPHICALLY PAIRED? CAN YOU MAKE AN ESTHETIC JUDGEMENT ON THESE THREE CLASSES?*

Table 8-1: ***Enumeration of Symmetry Configurations in the Plane.***

1, $1m$, k, $\underline{k}$, ∞, $\infty mm'$, ∞m and ∞g
$1\infty\infty'$, $1\infty\infty' mm'$, $1\infty\infty' mg$, $1\infty\infty' gg'$
$22'\infty$, $\underline{22'\infty}$, $2\tilde{2}\infty$
236, $\underline{236}$
$244'$, $\underline{244'}$, $\underline{2}4\tilde{4}$
$33'3''$, $\underline{33'3''}$, $\underline{3'}3\tilde{3}$
$22'2''2'''$, $\underline{22'2''2'''}$, $2\tilde{2}\underline{2'2''}$, $2\tilde{2}2'\tilde{2}'m/g$, and $2\tilde{2}2'\tilde{2}'g/g'$

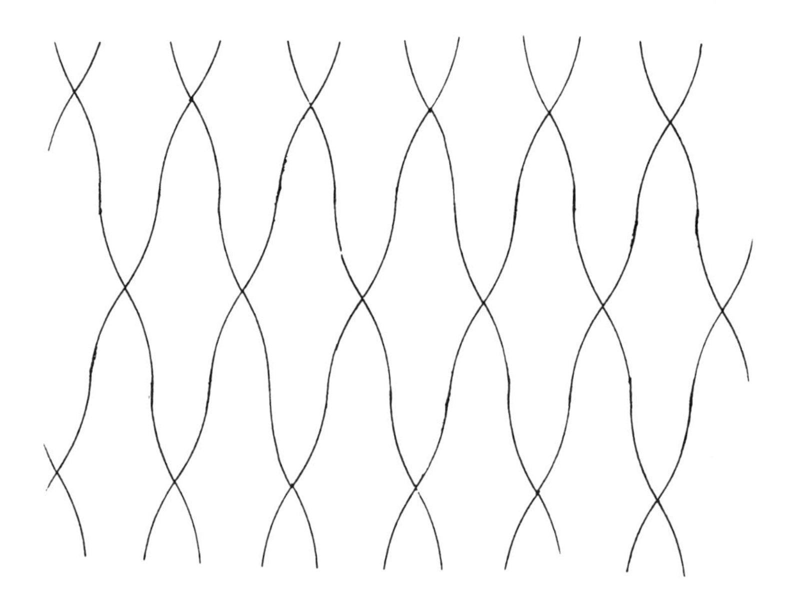

Figure 8-12 ***A pattern having the configuration*** $2\tilde{2}\underline{2'2''}$

Patterns having no enantiomorphy are almost too dynamic: Mirror symmetry gives a certain balance. On the other hand, patterns in which all rotocenters lie on mirrors tend to be rather static. Although the 236 is very attractive, it has the drawback that it does not allow enantiomorphically paired rotocenters. The $2\tilde{2}2'\tilde{2}'g/g'$ configuration is interesting because all rotocenters are enantiomorphically paired, and there are no mirror lines. The two configurations $\underline{2}4\tilde{4}$ and $\underline{3'}3\tilde{3}$ present a pleasing balance of rotocenters off mirrors, but paired enantiomorphically.

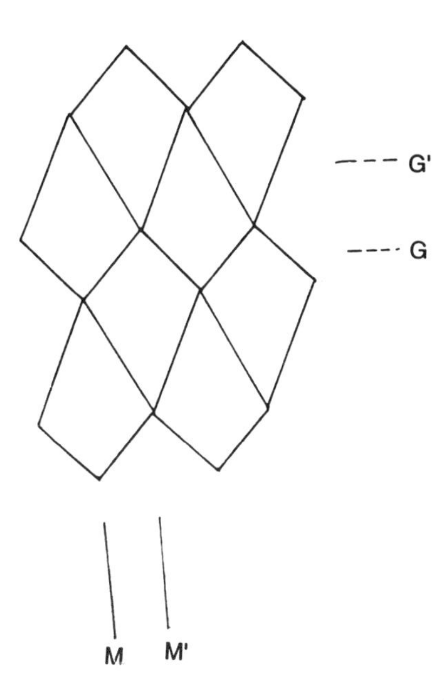

Figure 8-13 (a) ***A pattern having the configuration*** $2\tilde{2}2'\tilde{2}'m/g$

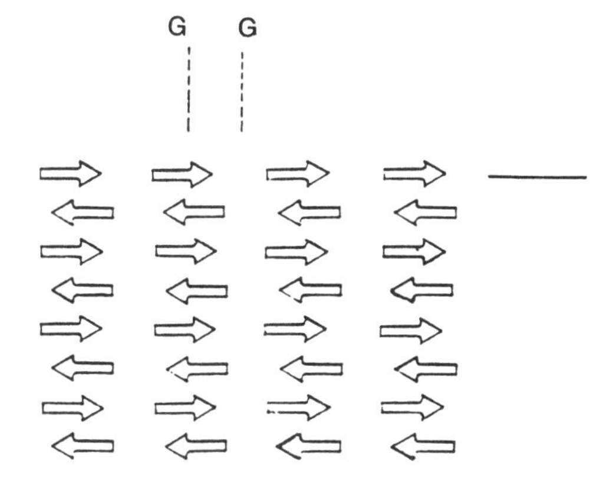

Figure 8-13 (b) ***Another illustration of the configuration*** $2\tilde{2}2'\tilde{2}'m/g$***: A one-way street grid pattern***

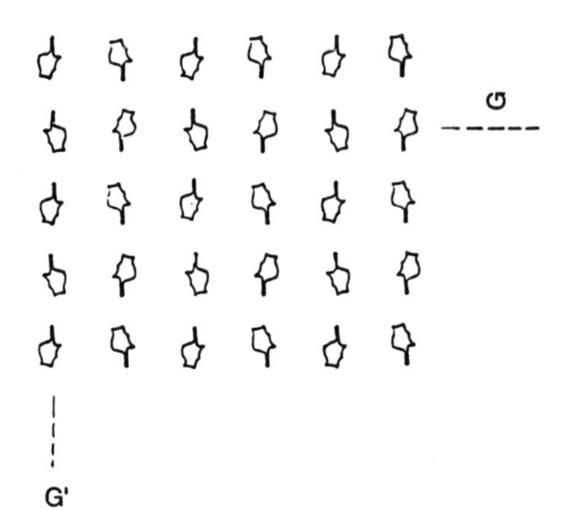

Figure 8-14 ***Illustration of the configuration*** $2\tilde{2}2'\tilde{2}'g/g'$

IX

Pentagonal Tessellations

In Chapter III we discovered that any straight-edged quadrilateral tessellates the plane; the demonstration that this is true introduced us to two-fold rotational symmetry. We demonstrated that curved-edged quadrilaterals whose edges are two-fold symmetrical also tessellate in the plane. Such tessellations have $22'2''2'''$ symmetry.

Q: *WHAT IS THE SYMMETRY OF THE QUADRILATERAL TESSELLATION IN FIGURE 3-18?*

The tessellation of Figure 3-18 has no finite rotational symmetry, but it has vertical alternating mirror and glide lines and diperiodic translational symmetry. Accordingly, its symmetry configuration is $1\infty\infty'mg$. In Chapters IV through VIII we made use of motifs to analyze the interactions of symmetry elements: Not all patterns are tessellations. However, tessellations may be designed having any of the diperiodic symmetry configurations.

Q: *DESIGN A TESSELLATION FOR EACH OF THE DIPERIODIC SYMMETRY CONFIGURATIONS.*

Figure 9-1 has another special quadrilateral tessellation, this one corresponding to symmetry configuration $\underline{3}'3\tilde{3}$. It was created by joining a rotocenter $\underline{3}'$ to a rotocenter 3 by a curved line; because of the mirror line on which $\underline{3}'$ lies by definition, there must be a second curved line joining $\underline{3}'$ to a rotocenter $3\tilde{3}$, this second line being the mirror image of the first one. Because of their rotational symmetry, each of 3 and $\tilde{3}$ are connected to three surrounding rotocenters $\underline{3}'$. The result is the quadrilateral tessellation of Figure 9-1, generated by a single curved line interacting with the symmetry configuration $\underline{3}'3\tilde{3}$. Note that three edges converge on the rotocenters 3 and $\tilde{3}$, while six edges converge on the rotocenters $\underline{3}'$.

Q: *CAN YOU DESIGN A STRAIGHT-EDGED PENTAGON WHICH TESSELLATES THE PLANE?*

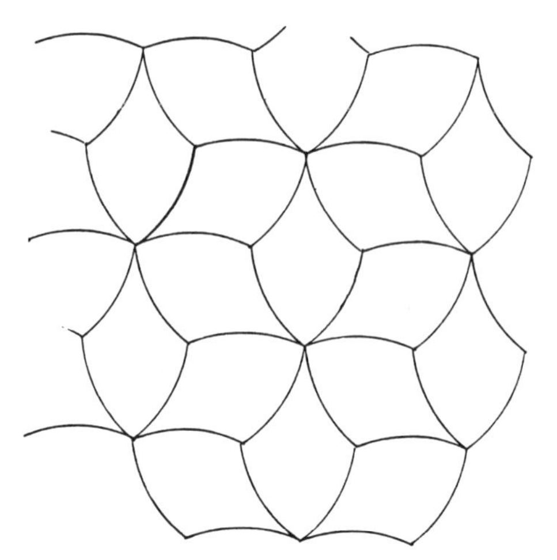

Figure 9-1 ***Quadrilateral tessellation having symmetry*** $\underline{3}'3\tilde{3}$

We discovered in Chapter VI that at most only a single five-fold rotocenter can exist in a plane. Since, moreover, the internal angle of a regular pentagon is 108°, it is not possible to fit an integral number of regular pentagons around a vertex. Therefore a regular pentagon does not tessellate the plane; as long as we know of at least one pentagon which does not tessellate the plane, we conclude that, while any straight-edged quadrilateral tessellates the plane, the same is not true of any straight-edged pentagon. Nevertheless, there are some *particular* pentagons which do tessellate the plane.

In Figure 9-2 we have drawn a pentagon which has two parallel edges AB and CD of equal length; a third edge joins ends A and C of these two lines together. An arbitrary point P is joined to B and D; the pentagon is ABPDC. A tessellation is to be effected having two-fold rotocenters halfway between P and B and halfway between P and D.

Q: *DRAW THE RESULTING PATTERN.*

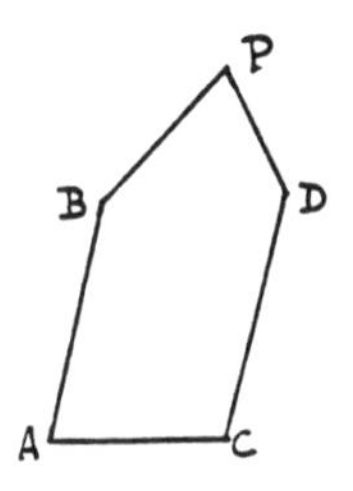

Figure 9-2 ***Pentagon tile***

Figure 9-3 shows the resulting pattern, a strip of symmetry $22'\infty$. The strip is bounded by two straight lines, one passing through points A and C. The plane can obviously be tessellated by covering it with adjacent strips equivalent to the one in Figure 9-3. There are two ways of doing so: The points A and C can be made into two-fold rotocenters with the result that all strips are mutually congruent. The symmetry of the resulting tessellation is $22'2''2'''$ (Figure 9-4). On the other hand, the line through A and C may be made into a mirror; adjoining strips will then be enantiomorphs, and the resulting tessellation has symmetry $2\tilde{2}2'\tilde{2}'m/g$ (Figure 9-5).

Q: *HOW WOULD CURVED-EDGED PENTAGONAL TESSELLATIONS LOOK HAVING THESE TWO SYMMETRY CONFIGURATIONS?*

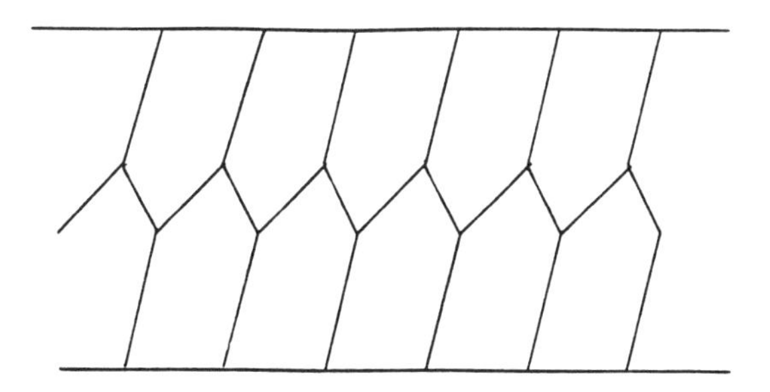

Figure 9-3 ***Monoperiodic pentagonal tessellation having symmetry*** $22'\infty$

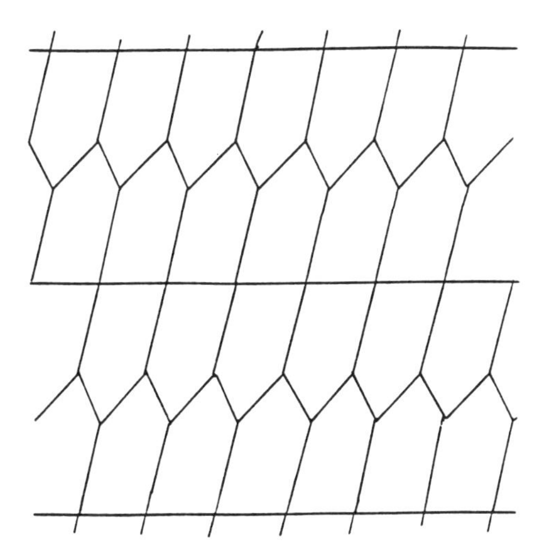

Figure 9-4 ***Pentagonal tessellation having symmetry*** $22'2''2'''$

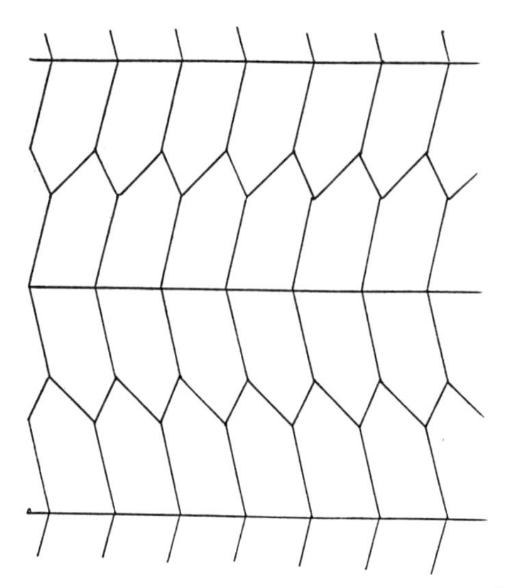

Figure 9-5 ***Pentagonal tesselation having symmetry*** $2\tilde{2}2'\tilde{2}'m/g$.

Figure 9-6 shows four very particular pentagons combined into a cross having four arms of equal length. Three angles of the pentagons are right angles; the remaining two angles are 135°. Two of the right angles are adjacent, and their sides are all of equal length. The third right angle lies between the 135° angles; its sides do not have the same length as do those of the other two right angles.

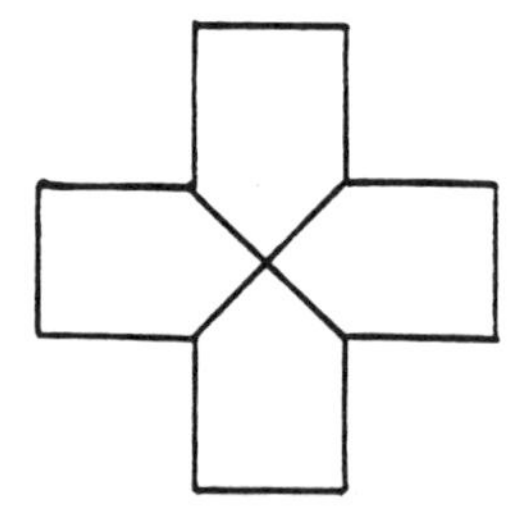

Figure 9-6 ***Cross composes of four special pentagons***

Q: *CAN YOU TESSELLATE THE PLANE WITH THIS CROSS?*

The cross composed of the four pentagons tessellates the plane. Although the cross has itself four lines of mirror symmetry, these mirrors cannot prevail in the tessellation, which has no mirror symmetry, hence it can exist in two mutually enantiomorphic manifestations (Figure 9-7 (p) and 9-7 (q). The symmetry configuration of both of these tessellations is $244'$.

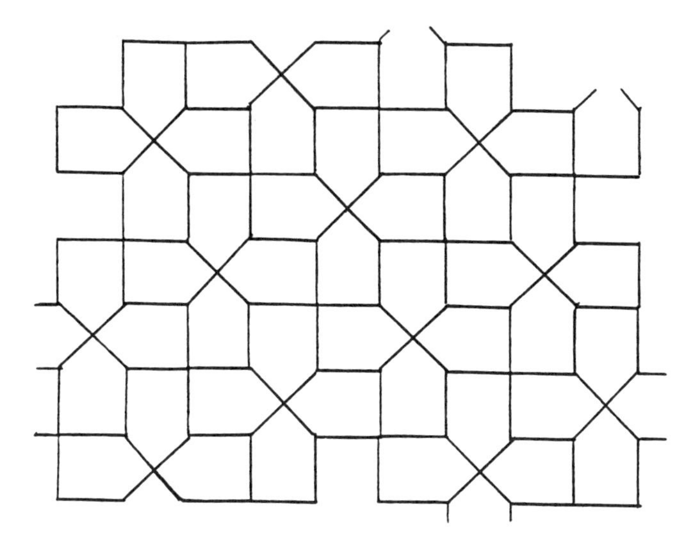

Figure 9-7 (p) ***Pentagonal tessellations using tiles of Figure 9-6***

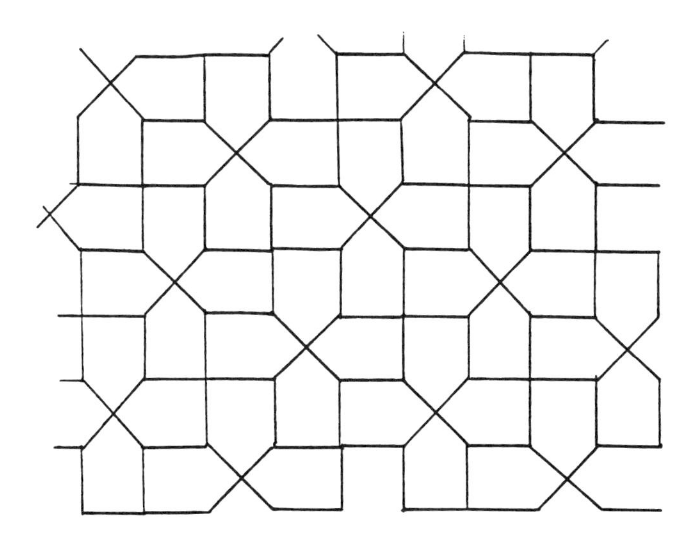

Figure 9-7 (q) ***Pentagonal tessellations using tiles of Figure 9-6***

Q: *WHY DO YOU SUPPOSE THESE ARE LABELED 9-7 (p) AND 9-7 (q) RATHER THAN 9-7 (a) AND 9-7 (b)?*

Q: *FIND THE TWO-FOLD ROTOCENTERS.*

Another tessellating pentagon is shown in Figure 9-8 (a). Two of its angles are right angles; they are separated by an angle whose magnitude is still to be determined, and which we shall call α. Four such pentagons meet around a right-angled vertex, which will be a four-fold rotocenter. A mirror line bisects angle α, and the other right-angled vertex will also be a four-fold rotocenter, enantiomorphic to the first one. Accordingly, the symmetry of the tessellations, shown in Figure 9-8 (b), is $\underline{2}4\tilde{4}$.

Q: *WHERE WOULD THE TWO-FOLD ROTOCENTERS BE?*

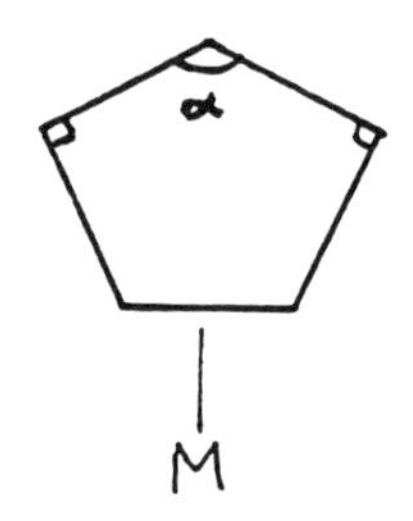

Figure 9-8 (a) ***Another special pentagon***

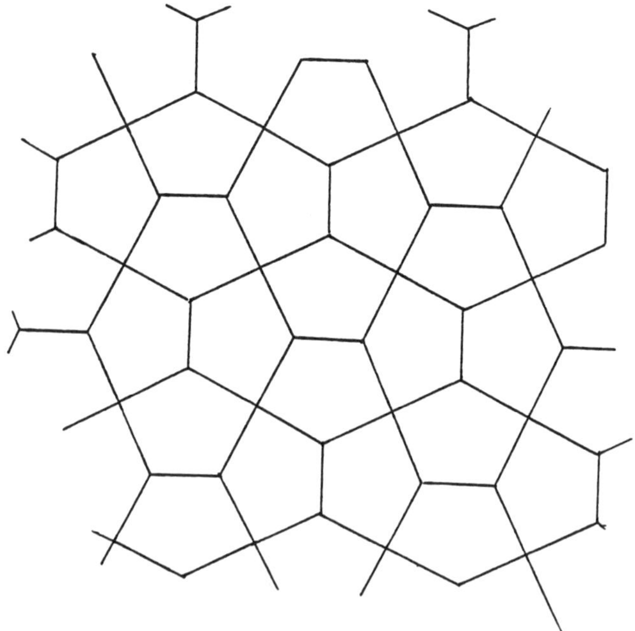

Figure 9-8 (b) ***Another special pentagonal tessellation***

The mesh of the 244′ system was shown in Figure 6-4. It is an isosceles right triangle whose vertices are a two-fold rotocenter at the right angle and two non-congruent four-fold rotocenters at the 45° angles. When the locations of the four-fold rotocenters are known, those of the two-fold rotocenters may be found by joining the nearest four-fold rotocenters by a straight line, which will be the hypotenuse of a mesh whose sides will make angles of 45° with this hypotenuse, and meet at a two-fold rotocenter. Two adjacent meshes are shown in Figure 9-9. The dotted lines are the edges of the pentagonal tiles inside one of the meshes. It is seen that, as remarked above, in $\underline{2}4\tilde{4}$ the meshes are bisected by mirrors, and two adjacent meshes contain different information.

The tessellation of Figure 9-8 (b) has symmetry $\underline{2}4\tilde{4}$.

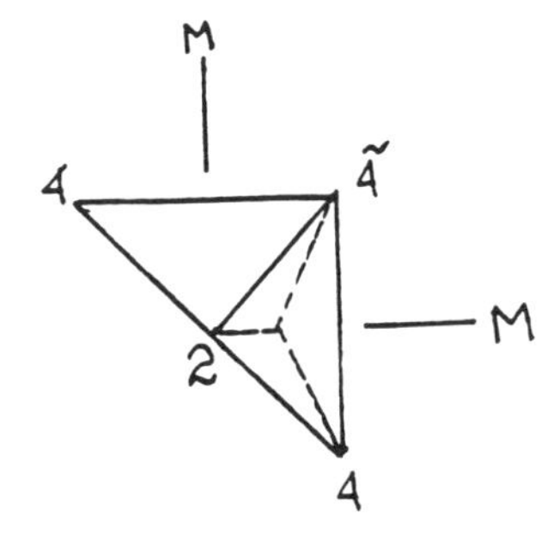

Figure 9-9 ***Two adjacent meshes of the tessellation of Figure 9-8 (a)***

Q: *WHAT IS THE VALUE OF ANGLE α? HOW ARE THE LENGTHS OF THE FIVE SIDES OF EACH PENTAGON RELATED?*

Angle α may vary all the way from a minimun of 90° to as much as 270°, as shown in Figures 9-10 (a)($\alpha = 90°$), 9-10 (b) ($\alpha = 120°$), 9-10 (c) ($\alpha = 180°$) and 9-10 (d) ($\alpha = 240°$). Figure 9-11 is a transformation: Starting at the top with the pentagonal tessellation of Figure 9-8 (b), we remove the cross bar passing through the two-fold rotocenters, producing a quadrilateral tessellation. Then, going down, we insert new crossbars, producing a tessellation having $\alpha > 180°$. Finally, at the bottom, we demonstrate that the pentagonal tessellation of Figure 9-8 (b) may be considered to be the superposition of two mutually congruent hexagonal tessellations oriented perpendicular to one another, these hexagons being irregular.

In the tessellations of Figure 9-8 through 9-11 four sides of the pentagon are of equal length, with the length of the "cross bar" passing through the two-fold rotocenters depending on the magnitude of the angle α.

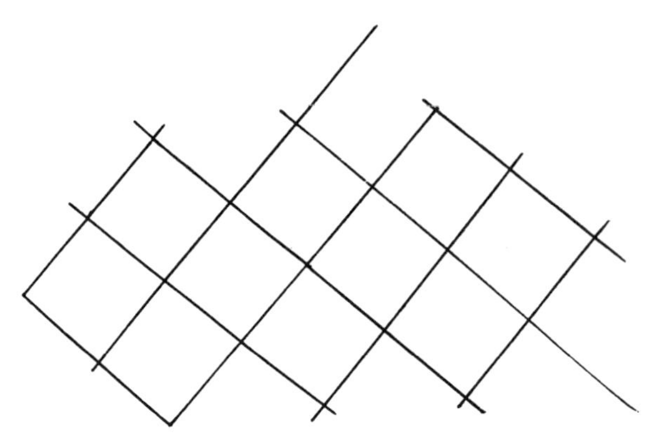

Figure 9-10 ***Tessellations for different values of*** α ***(a):*** $\alpha = 90°$

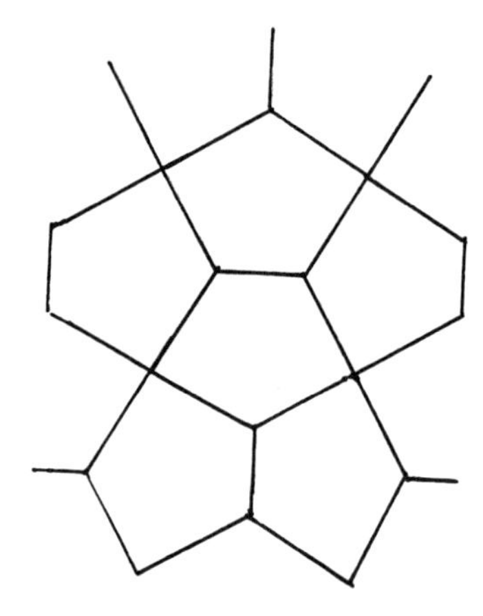

Figure 9-10 (b) $\alpha = 120°$

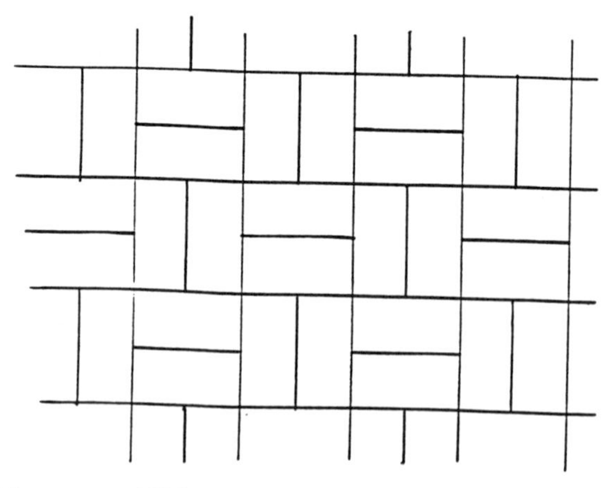

Figure 9-10 (c) $\alpha = 180°$

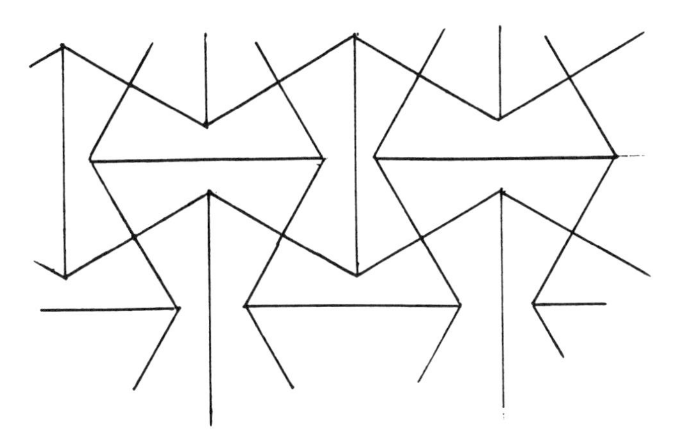

Figure 9-10 (d) $\alpha = 240^\circ$

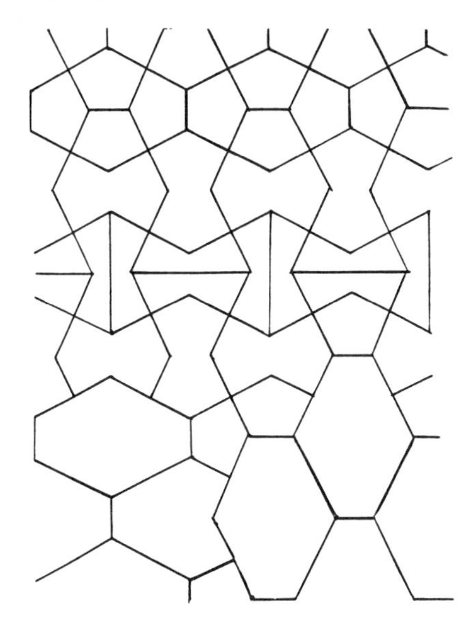

Figure 9-11 ***Transformation of Figure 9-10 (b)***

Figure 9-12 shows a pentagonal tile in which each of the right angles is surrounded by edges of equal length, but the lengths of the edges surrounding one of the right angles is different from those surrounding the other right angle.

Q: *WOULD SUCH A PENTAGON TILE THE PLANE?*

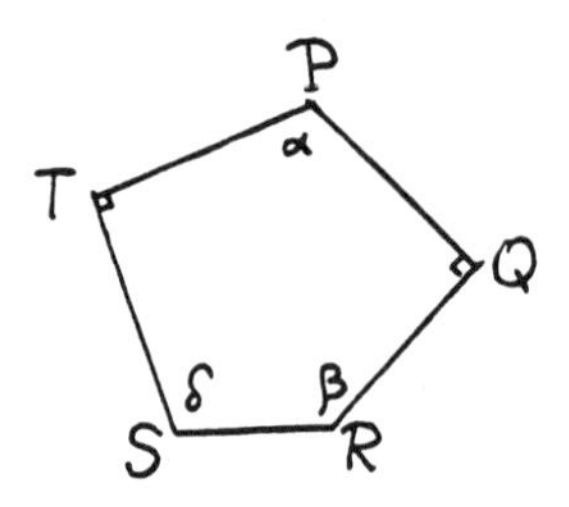

Figure 9-12 ***Skewed Pentagonal Tile***

Label the vertices of the pentagon as follows: The vertex adjacent to both right-angled vertices is called P; moving clockwise, the vertex adjacent to P is called Q, and so on. The right-angled vertices are accordingly labeled Q and T. Call the angle QPT α, angle QRS β, angle RST δ. Rotate the pentagon 180° around point C at the center of edge RS, and also 90° around both vertex Q and vertex T.

Q: *CONTINUE TESSELLATING. WHAT IS THE SUM OF ANGLES β AND δ?* (Remember that the sum of the angles of a pentagon is 540°.)

Figure 9-13 (a) shows pentagon PQRST together with P′Q′SRT′, the result of rotating PQRST around C, the center of the line segment RS. A path from Q to T along edges of the first pentagon has two components: QP and PT. A path from Q′ to T similarly has two components, Q′S and ST. From the definition of pentagon PQRST, PT and TS are mutually perpendicular and equal in length. Because of the rotational symmetry, PQ is parallel to and equal in length to P′Q′, which by the definition of pentagon P′Q′SRT′ is perpendicular to and equal in length to Q′S. As a result, QT is perpendicular to and equal in length to QT′; by a similar argument we can show that Q′T′ is perpendicular to and equal in length to QT′ and Q′T. Therefore QTQ′T′ is a square, whose diagonals perpendicularly bisect each other at C, the center of RS. Figure 9-13 (b) shows a third pentagon, resulting from a 90° rotation of PQRST around the point T. It demonstrates that PQRST does indeed tessellate the plane, with generally distinct four-fold rotocenters at the right angles, and a two-fold rotocenter at the center of RS.

Q: *TRACE THESE THREE PENTAGONS ON TRANSPARENT PAPER, AND COMPLETE THE TESSELLATION. INDICATE THE LOCATION OF THE ROTOCENTERS AND A PAIR OF ADJACENT MESHES.*

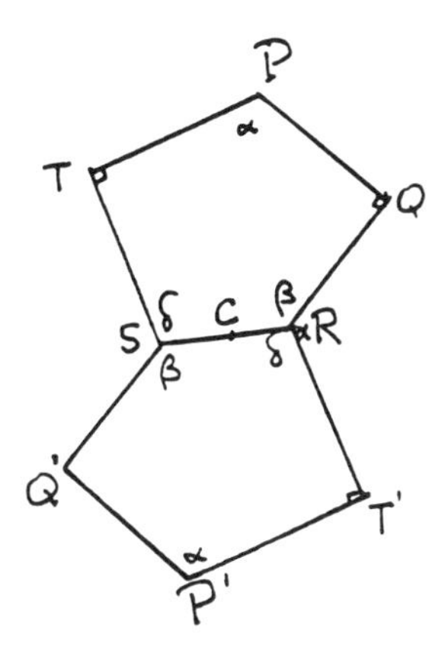

Figure 9-13 (a) ***A pair of mutually congruent tiles***

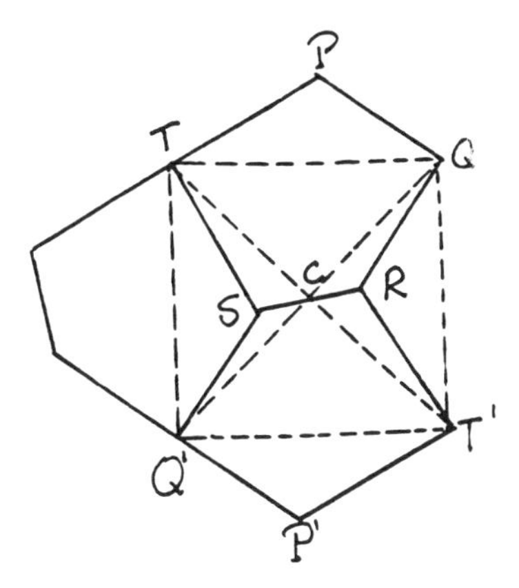

Figure 9-13 (b) ***Three mutually congruent tiles***

The triangle QCT is a mesh of the 244′ system. That the pentagon PQRST tessellates the plane is possible because QTQ′T′ is a square; its diagonals perpendicularly bisect each other at C. Therefore, the pentagon PQRST has the following peculiar property:

Theorem 9-1: A pentagon which has two non-adjacent right angles, each of which has equal sides, has a fifth side whose center has the following properties: It is equidistant from the two right-angled vertices, and it subtends a right angle at those vertices.

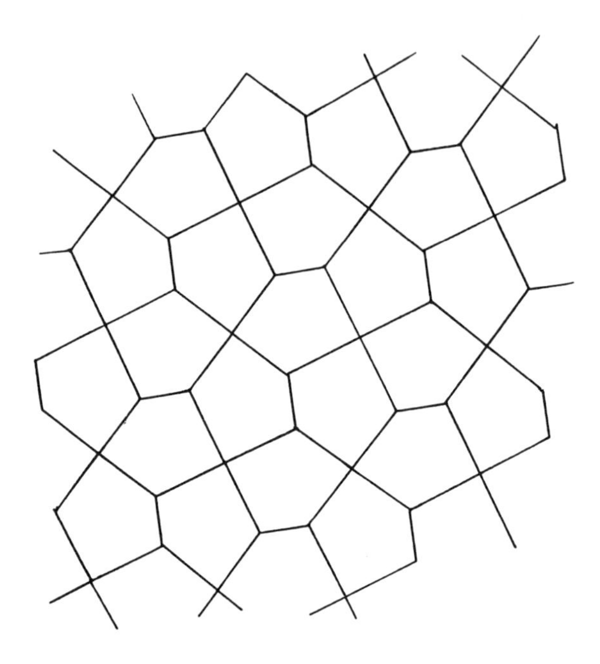

Figure 9-14 ***Tessellation using a skewed pentagon***

Q: *FIND THE SPECIAL CASE FOR THE TESSELLATION IN FIGURE 9-14 FOR WHICH* $\beta = 90°$, *AND* $\alpha = \delta$.

Since $\alpha + \beta + \delta = 360°$, if $\alpha = \delta$ and $\beta = 90°$, then $\alpha = \delta = 135°$. Therefore the tessellations shown in Figure 9-7 are actually special cases of the tessellation commenced in Figure 9-13 (b). Of course, those of Figure 9-10 are also special cases of that tessellation.

Those who want to know more about tessellating pentagons are referred to the work of Marjorie Rice.[1]

Q: *DESIGN SOME CURVED-EDGED PENTAGONAL TESSELLATIONS.*

Notes

[1] Schattschneider, Doris: *Tiling the Plane with Congruent Pentagons*, Math Magazine, **51**, 20–44 (1980).

X

Hexagonal Tessellations

Figure 10-1 shows how a pentagonal tessellation may be considered as the superposition of two mutually perpendicular and congruent hexagonal tessellations. Figure 10-2 shows a hexagonal tessellation in which pairs of opposite edges of each tile are mutually parallel and of equal length. The angles α and β occur twice in each hexagon; since the angles of a hexagon add up to 720°, the two remaining angles are $360° - \alpha - \beta$.

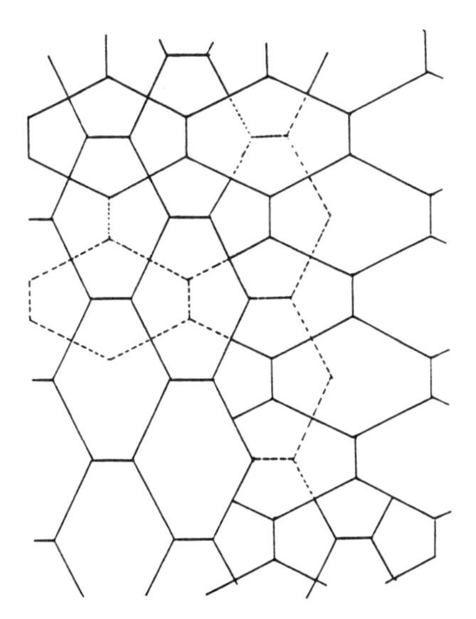

Figure 10-1 ***Two hexagonal tessellations superimposed to produce a pentagonal tessellation***

Q: *WHAT IS THE SYMMETRY OF THE HEXAGONAL TESSELLATION OF FIGURE 10-2? DOES IT HAVE ROTATIONAL SYMMETRY? REFLECTION SYMMETRY?*

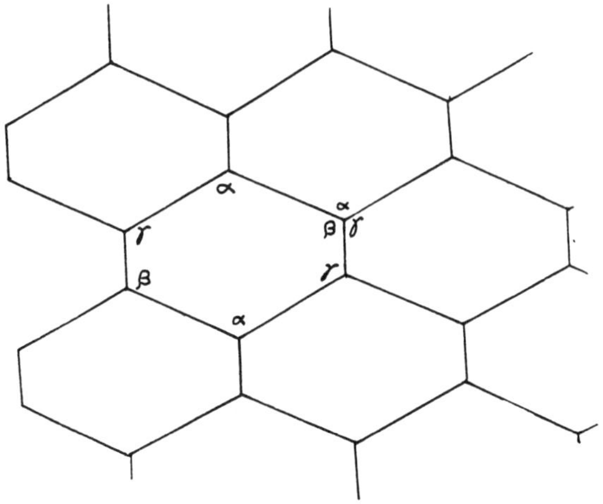

Figure 10-2 ***Hexagonal tessellation***

Each hexagon in the tessellation of Figure 10-2 has mutually equivalent two-fold rotocenters at the center of each pair of opposite edges. A fourth rotocomplex occurs at the intersection of the diagonals of each hexagon; this "central" rotocenter relates the mutually equivalent rotocenters on opposite edges to each other. If $\alpha = \beta = 120°$ and all edges have equal length, then Figure 10-2 turns into a tessellation of regular hexagons.

Q: *WHAT IS THE SYMMETRY OF A REGULAR HEXAGONAL TESSELLATION?*

Q: *IS THE TESSELLATION OF FIGURE 10-2 THE MOST GENERAL HEXAGONAL TESSELLATION POSSIBLE?*

To consider this question, let us weave a hexagonal tessellation through a $22'2''2'''$ system. Remember that there may not be more than *four* distinct rotocenters. Four of the six edges of the hexagon will pass through two-fold rotocenters, hence have that symmetry themselves (cf. Figure 10-3). The remaining edges form "bridges" between strings of edges each having $22'\infty$ symmetry. Observe that these curved bridges are mutually related by two-fold rotational symmetry, hence half of these bridges curve in one direction, the others in the opposite direction.

Q: *LOOK AT THE EDGES ON OPPOSITE SIDES OF ANY GIVEN HEXAGON. HOW ARE THEY RELATED?*

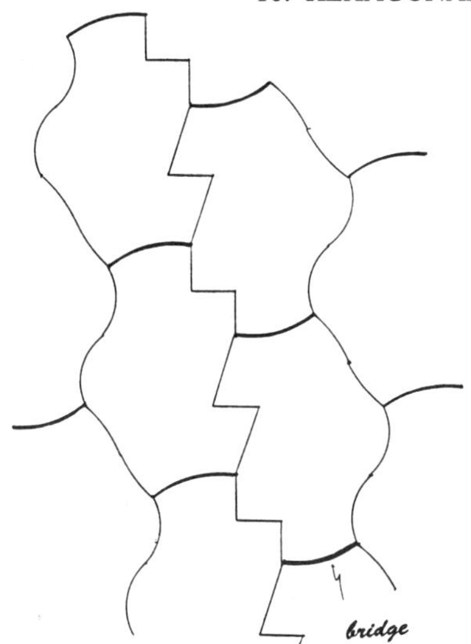

Figure 10-3 ***Curved-edged hexagonal tessellation***

Each hexagon of Figure 10-3 is characterized by four edges having two-fold rotational symmetry. The remaining edges are related to each other by two successive 180° rotations, that is to say, by translation. In order to avoid violating the postulate of closest approach, the centers of the two-fold symmetrical edges must constitute the vertices of a parallelogram. A hexagon which tessellates the plane by means of 180° rotations needs to satisfy these characteristics. In Figure 10-4 is shown a hexagon having a pair of opposite edges, PQ and TS, related by translational symmetry. Two points, R and U have been chosen arbitrarily as additional vertices, and these are joined respectively to P and T and to Q and S by means of two-fold symmetrical edges.

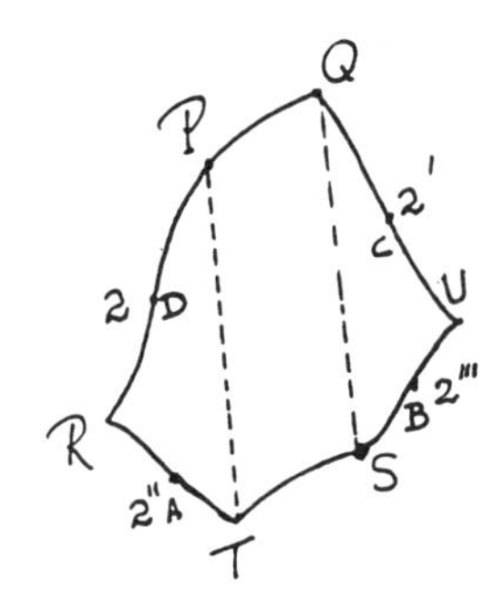

Figure 10-4 ***A hexagon which tessellates the plane***

Q: *DO THE TWO-FOLD ROTOCENTERS ON THESE EDGES PR, RT, QU AND US NECESSARILY LIE ON THE VERTICES OF A PARALLELOGRAM?*

If so, then the hexagon of Figure 10-4 will assuredly tessellate the plane, but if not, then some additional restraint needs to be formulated for a tessellating hexagon.

Draw straight lines in Figure 10-4 from P to T, and from Q to S. Because of the translational symmetry between PQ and TS the lines PT and QS are mutually parallel and equal in length. Label the center of RT, A, that of SU, B, that of QU, C, and that of PR, D. The line DA joins the centers of sides of the triangle RPT, hence is parallel to and half as long as the base of that triangle, PT. Similarly, the line CB is parallel to and half as long as the line QS. Since the lines PT and QS are mutually parallel and equal in length, the lines DA and CB are also equal in length and parallel, so that A, B, C and D indeed lie at the vertices of a parallelogram, and the only restrictions on a tessellating hexagon are that a pair of opposite edges are related by translational symmetry, and the remaining edges are two-fold symmetrical.

Interesting is the tile of Figure 10-5. At first sight this figure would not appear to have four edges having two-fold rotational symmetry. However, as Figure 10-6 shows, two-fold rotocenters are indeed present, so that the tessellation condition is indeed satisfied.

Figure 10-5 ***Will this hexagon tessellate the plane?***

Figure 10-6 ***Tessellation using the tile of Figure 10-5***

Doris Schattschneider[1] cautions us about translational symmetry by means of a figure like 10-7: Here PQ is parallel to ST, not to TS as we discussed above. It is important to understand translational symmetry in this respect.

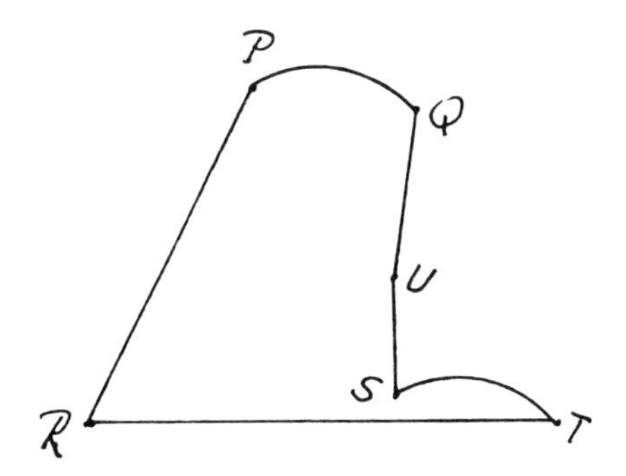

Figure 10-7 ***Schattschneider's cautionary warning***

NOTES

[1] Schattschneider, Doris: *Tiling the Plane with Congruent Pentagons*, Math Magazine, **51**, 29–44 (1980).

XI

Dirichlet Domains

If you live in a city having subway service, it would be useful to be able to read from a map which subway station would be closest to any location where you may find yourself at a given moment. And if you go to college, you might want to know at any moment where the nearest dining hall would be. If you have children, you might want to know the location of the nearest school. Churches, cities, any center of attraction, actually have their own regions, defined such that any location within that region is closer to the center within its borders than to any other center. Such a region is called a Dirichlet domain, after a mathematician whose wife, incidently, was a sister of composer Felix Mendelssohn. Dirichlet domains are regions associated with arrays of discrete points.

To learn how to construct Dirichlet domains, consider first just a pair of points, P and Q, as shown in Figure 11-1. The entire plane may be divided into two domains such that every location in one is closer to P, and in the other is closer to Q. The boundary between these two regions is the perpendicular bisector of the line PQ. That bisector is the locus of all points equidistant from P to Q. [A *locus* is an array of all points that satisfy a given condition: A circle is the locus of all points equidistant from a given point.]

Q: *SUPPOSE THAT YOU HAD THREE POINTS P, Q AND R LYING ON A COMMON STRAIGHT LINE. WHAT WOULD THEIR DIRICHLET DOMAINS LOOK LIKE? WHICH WOULD APPEAR TO HAVE THE SMALLEST DIRICHLET DOMAIN?*

In Figure 11-2 we have drawn three points P, Q and R which do not all lie on the same straight line. The boundary lines of the Dirichlet domains are portions of the perpendicular bisectors of the three lines PQ, QR and RP.

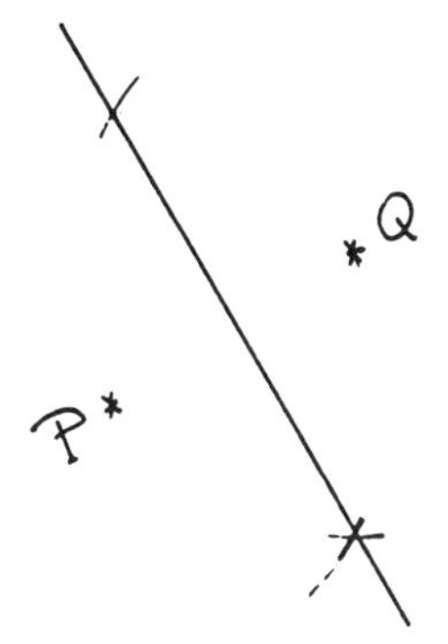

Figure 11-1 ***Dirichlet domains of a pair of points***

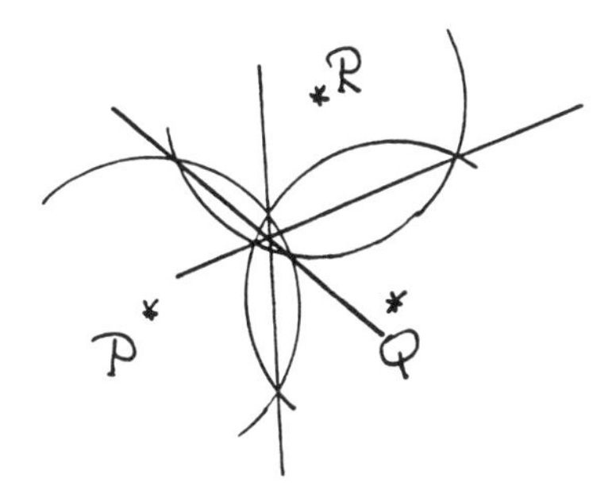

Figure 11-2 ***Dirichlet domains of three non-collinear points***

Q: *IN FIGURE 11-2 THESE THREE BISECTORS MEET AT A COMMON POINT. WILL THIS ALWAYS BE THE CASE?*

All points on the perpendicular bisector of PQ are equidistant from P and Q. All points on the perpendicular bisector of QR are equidistant from Q and R. The point at the intersection of these two bisectors is therefore equidistant from P, Q and R, hence lies on the perpendicular bisector of RP. Therefore all three perpendicular bisectors have a common point of intersection.

Q: *IN FIGURE 11-2, PQR IS A TRIANGLE ALL OF WHOSE ANGLES ARE ACUTE. DRAW THE DIRICHLET DOMAINS OF POINTS AT THE VERTICES OF A TRIANGLE WHICH HAS ONE OBTUSE ANGLE.*

In Figure 11-3 four points, P, Q, R and S, are drawn at the vertices of a quadrilateral.

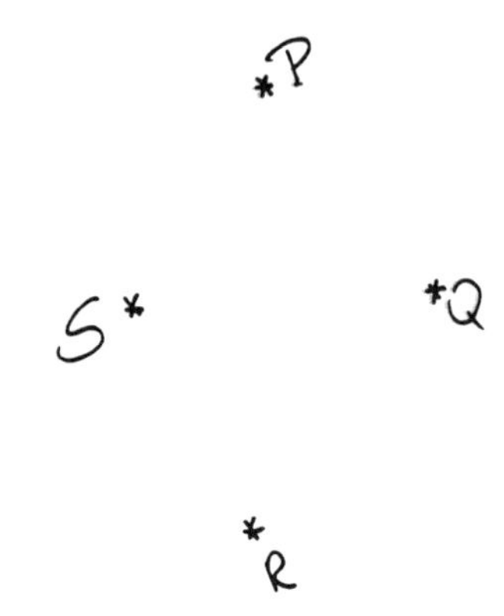

Figure 11-3 ***Four points***

Q: *HOW MANY LINES WOULD HAVE TO BE PERPENDICULARLY BISECTED?*

If there are, in general, N points which have to be connected, then every one of these will need to be connected to $(N-1)$ other points. That would appear to imply that there would in total be $N(N-1)$ connections, but we must realize that every connection joins two points, hence was counted twice. Therefore the number of connections between N points would be $\frac{1}{2}N(N-1)$. For $N = 3$ there are three connections, for four points the number of connections is six. For 100 points the number of connections is 4950. Obviously, as the number of centers increases, a veritable tangle of bisectors would be generated. Even in the case of four points, generally not all six perpendicular bisectors will pass through a single point.

There is a method, however, which will help untangle these connections, and that is to take near points three at a time. You may have noted that for a triangle which has an obtuse angle, the perpendicular bisectors meet outside that triangle. Such a configuration should therefore be avoided if possible. In Figure 11-4 we drew a diagonal dividing the quadrilateral of Figure 11-3 into two acute triangles; each separate triangle was then treated as triangle PQR was in Figure 11-2. We note that each point has indeed acquired a Dirichlet domain within which every location is closer to that point than to any of the other three. We also note that every one of the four points shares Dirichlet domain boundaries with at least two other points. Of these four points, two share boundaries with each of the other three points, but the other two share boundaries with two others only.

We shall use the concept of Dirichlet domains to define *neighbors* among an array of points: Neighbors are those points whose Dirichlet domains share a finite boundary. Points whose Dirichlet domains touch at a single point only are not considered neighbors. Among the points P, Q, R and S in Figure 11-3, there are five pairs of neighbors; the sixth pair are not neighbors (cf. Figure 11-4).

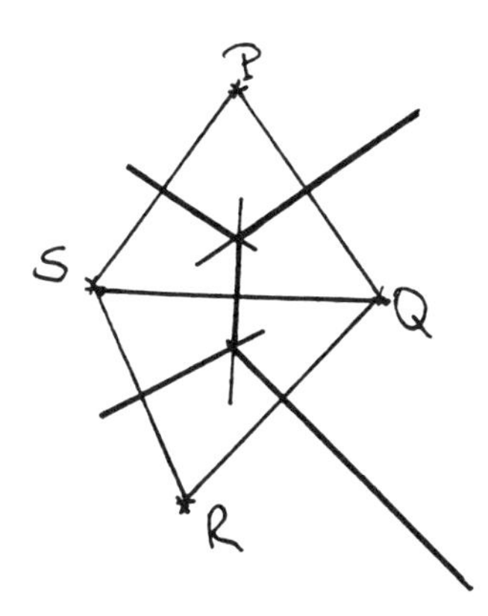

Figure 11-4 ***Dirichlet domains of four points***

Q: *COULD YOU DRAW AN ARRAY OF FOUR POINTS ALL OF WHICH ARE NEIGHBORS? OR ONE IN WHICH EVERY POINT HAS ONLY TWO NEIGHBORS?*

In the configuration of Figure 11-5 obtuse triangles cannot be avoided. Here all points are neighbors; one Dirichlet Domain is completely bounded, the other three extend indefinitely. Note that in both Figures 11-4 and 11-5 not more than three boundaries meet in any point.

In exceptional situations more than three perpendicular bisectors may meet at a common point. This meeting point would be equidistant from more than three points; this means that the meeting point is the center of a circle on whose circumference are more than three domain centers.

If points P, Q, R and S in Figure 11-4 had been located on the circumference of a common circle, their Dirichlet domains would all meet at a point (Figure 11-6). In this case each of the four points has only two neighbors, because domains which meet in a point without sharing a finite boundary are not considered neighbors.

In constructing Dirichlet domains one should always make certain that every location within a designed domain is indeed closer to its center than

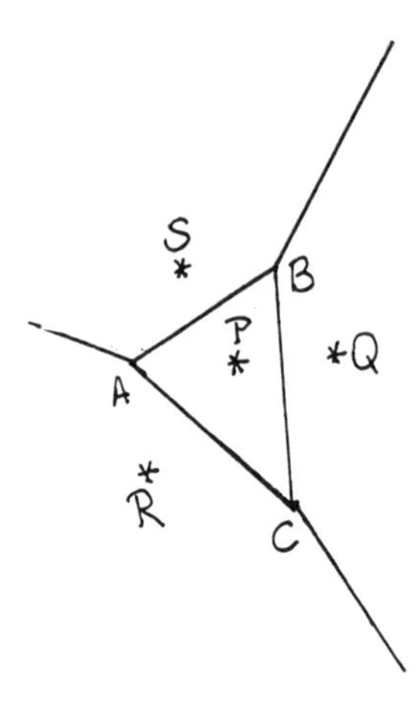

Figure 11-5 ***All four points are neighbors of the other three***

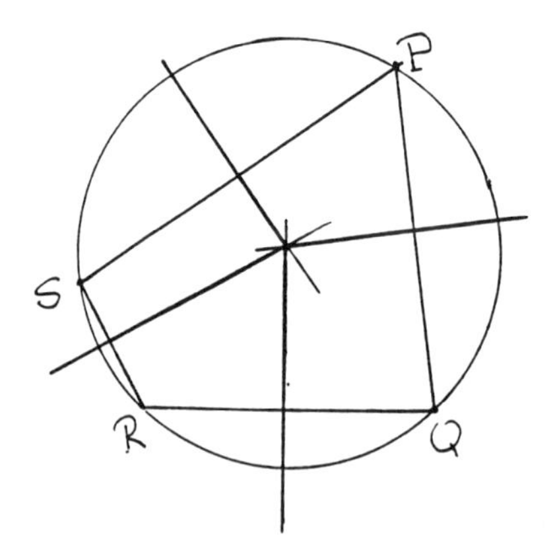

Figure 11-6 ***Dirichlet domains of four points lying on the circumference of a common circle***

to any other. The triangulation method, if applied mechanically, will not always produce true Dirichlet domains and must therefore be used with caution and good judgement. In particular, corners jutting out into or between adjacent domains must be viewed skeptically, for they should probably be truncated by bisectors which had not been extended sufficiently far.

Our chief interest in the context of this book, however, are Dirichlet domains of regularly spaced points, such as shown in Figure 11-7. Such points, being in identical contexts, will have identical Dirichlet domains. In a plane which is populated by a set of discrete points, every location will either be closer to one of these points than to any other or will be equidistant from two or more of the points. In the former case it will lie inside a Dirichlet domain, in the latter on a domain boundary. In any case, there is no point in the plane not either within, or on the boundary of at least one Dirichlet domain; therefore Dirichlet domains together fill the plane without overlap or interstice. The Dirichlet domains of regularly spaced points therefore tessellate the plane.

Q: *DRAW THE DIRICHLET DOMAINS FOR THE ARRAY SHOWN IN FIGURE 11-7. HOW MANY NEIGHBORS DOES EACH POINT HAVE?*

Figure 11-8 shows an array of points located at the vertices of tesselating parallelograms.

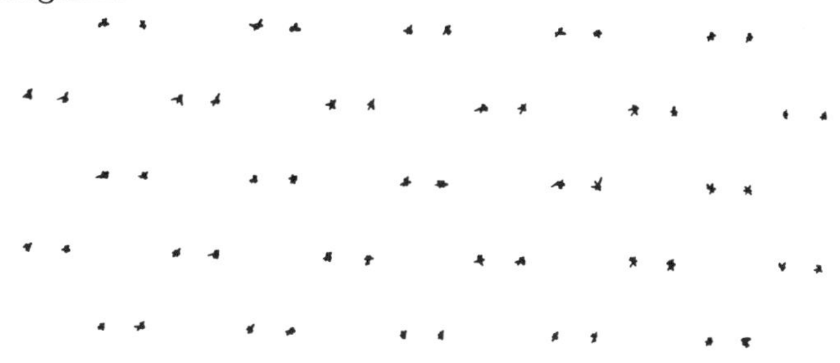

Figure 11-7 ***An array of regularly spaced points***

Figure 11-8 ***Another array of regularly spaced points***

Q: *WITHOUT DRAWING THE DIRICHLET DOMAINS, HOW MANY NEIGHBORS WOULD YOU ESTIMATE EACH POINT TO HAVE?*

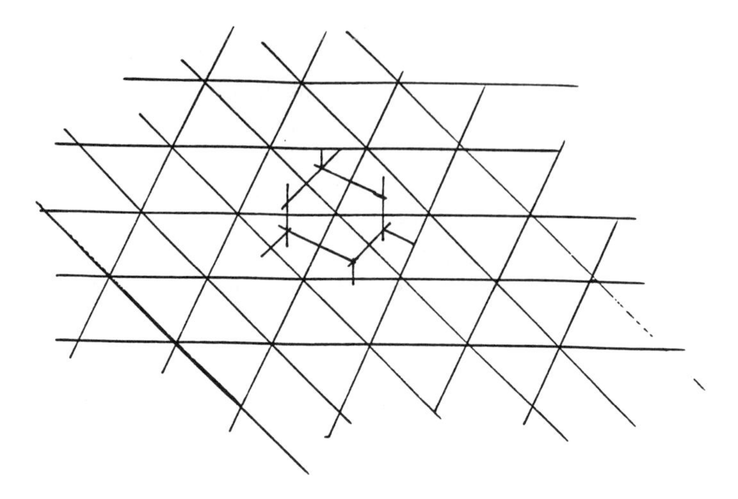

Figure 11-9 ***Triangulation and Dirichlet domains of Figure 11-8***

Figure 11-9 triangulates the array, and a Dirichlet domain is drawn. It is a hexagon; therefore every point has six neighbors.

Q: *DOES THIS HEXAGON CONFORM TO THE CRITERIA FOUND IN CHAPTER X FOR TESSELLATING WITH HEXAGONS?*

In Figure 11-10 the parallelograms of Figure 11-8 are transformed into rectangles. Note that two of the sides of the hexagons have vanished, so that the Dirichlet domains now have become rectangles as well, with the result that each point here only has four neighbors.

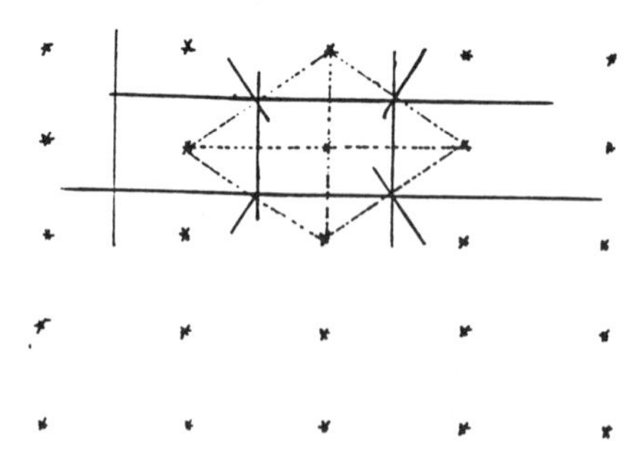

Figure 11-10 ***Rectangular array of points***

Q: *SINCE FOUR DIRICHLET DOMAINS MEET AT A COMMON VERTEX, ONE WOULD EXPECT THE CENTERS OF SUCH DOMAINS TO LIE ON A COMMON CIRCLE. IS THIS TRUE HERE?*

The vertices of the Dirichlet domains constitute a regularly spaced array of discrete points, which in turn would have their own Dirichlet domains.

Q: *DRAW THESE DOMAINS AROUND THE VERTICES OF THE DOMAINS OF THE ARRAY OF FIGURE 11-8.*

Observe that the vertices of these latter domains are located precisely on the original array of discrete points! The two patterns of Dirichlet domains bear an interesting relationship to each other, which we shall examine further, and which is called *duality*. We must caution, however, that in three dimensions these relationships are more complex, and should not be extrapolated from two dimensions without care.[1]

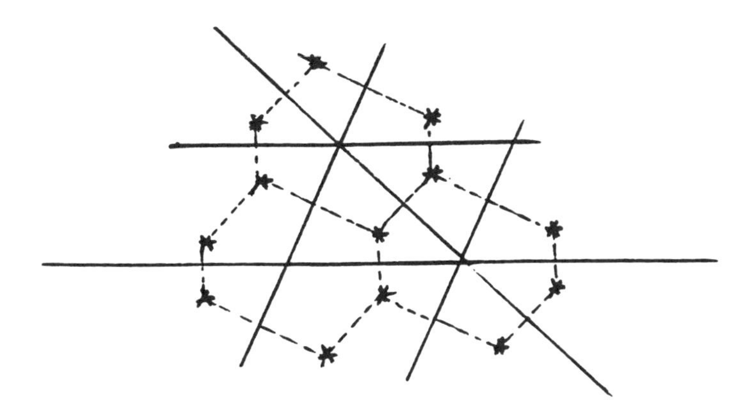

Figure 11-11 ***Dirichlet domains of the vertices of the Dirichlet domains of Figure 11-8***

We noted that the points in Figure 11-8 have six neighbors, for their Dirichlet domains are hexagons. The vertices of these hexagonal domains have triangular Dirichlet domains, hence three neighbors. The term *valency* may be applied to a point (vertex) as well as to a polygon (face): The valency of a vertex is the number of edges joined at the vertex, whereas the valency of a face is the number of edges enclosing that face. The hexagonal domains of Figure 11-8 accordingly are hexavalent, as are the vertices of the triangular domains of Figure 11-11. Conversely, the domains in Figure 11-11 are trivalent, as are the vertices of the Dirichlet domains in Figure 11-9. The vertices in Figure 11-9 correspond to the faces in Figure 11-11; the faces in Figure 11-9 correspond to the vertices in Figure 11-11. These two figures are called ***duals***: Dual figures are those in which the vertices of one correspond to the faces of the other, and vice versa. When the two dual figures are superimposed, the edges of the Dirichlet domains of one join neighbors in the other, and vice versa (Figure 11-12). Duality is a general property not restricted to the Dirichlet domains discussed here. There are tessellations which are not Dirichlet domains; the Dirichlet tessellations are very special examples of tessellations.[2]

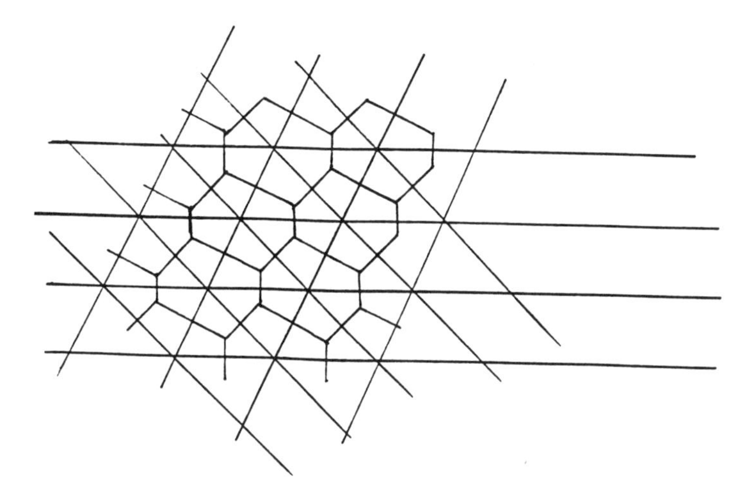

Figure 11-12 ***Duality***

Note that the edges of duals cross each other: When duals are superimposed, every edge in one figure is crossed by an edge in the other. Every vertex in one figure lies in a face of the other, and every face in one contains a vertex in the other. This is the specific meaning of the term "corresponds" in the definition of duality. The vertices of one do not need to lie in any specific location in the face of the other, and the edges do not need to cross at any particular angle. However, since the construction of Dirichlet domains involves ***perpendicular*** bisection, the edges of Dirichlet duals will perpendiculalry bisect each other.

Now turn to Figure 11-10, and its Dirichlet domains.

Q: *WHAT IS THE VALENCY OF THE DIRICHLET DOMAINS? OF THEIR VERTICES?*

The Dirichlet domains of this array of points constitute a rectangular tessellation, conversely, the vertices in that tessellation also have rectangular domains. In a rectangular tessellation all faces as well as all vertices are quadrivalent. The dual figure of the rectangular tessellation is the identical rectangular tessellation. A configuration which is identical to its own dual is called *self-dual*. A rectangular tessellation is self-dual.

Dirichlet domains thus constitute a special class of tessellating polygons associated with an array of discrete points, whose vertices have their own Dirichlet domains. The vertices of the latter domains constitute the original array of discrete points. The two Dirichlet tilings are dually related.

NOTES

[1] Loeb, A. L.: *Space Structures, Their Harmony and Counterpoint,* Birkhäuser Boston/Berlin/Basel, 132 (reprinted 1991).

[2] Loeb, A. L.: op. cit., Chapter 14.

XII

Points and Regions

In this chapter we summarize the various arrays of discrete points and the various tessellating polygons which we have encountered in the preceding chapters, and introduce some others. Notably, there is a one-to-one correspondence between some of the discrete points and the polygons.

IT WILL BE CONVENIENT TO HAVE A SUPPLY OF TRACING PAPER HANDY FOR THIS CHAPTER.

A *lattice complex* is the array of all discrete points related symmetrically, that is to say, whose contexts are identical, even though oriented differently, and possibly enantiomorphic. Every point in the plane belongs to some lattice complex, being symmetrically related to every other point belonging to that same lattice complex.

Q: *IN FIGURE 12-1 WE SEE A PATTERN OF SYMMETRICALLY RELATED STAR FIGURES. IN THIS PLANE A POINT IS MARKED P. TAKE A PIECE OF TRACING PAPER, AND MARK WITH ASTERISKS (*) THE POINT P TOGETHER WITH ALL OTHER POINTS SYMMETRICALLY RELATED TO P. CALL THIS FIGURE 12-1 (a).*

Associated with a lattice complex is the *Dirichlet domain*, the special tessellating polygon discussed in Chapter XI.

Q: *DRAW A FEW DIRICHLET DOMAINS FOR THE POINTS WHICH YOU MARKED ON YOUR TRACING PAPER.*[1]

Rotocomplexes constitute special lattice complexes: The points constituting a rotocomplex all lie on rotocenters. A single rotocenter is a complex having a single member. As we showed in Chapter V, the coexistence of two rotocenters in a plane implies an infinite array of rotocenters, whose symmetry values are limited by a diophantine equation. The combinations of symmetry values permitted were found to be $1\infty\infty'$, $22'\infty$, 236, 244',

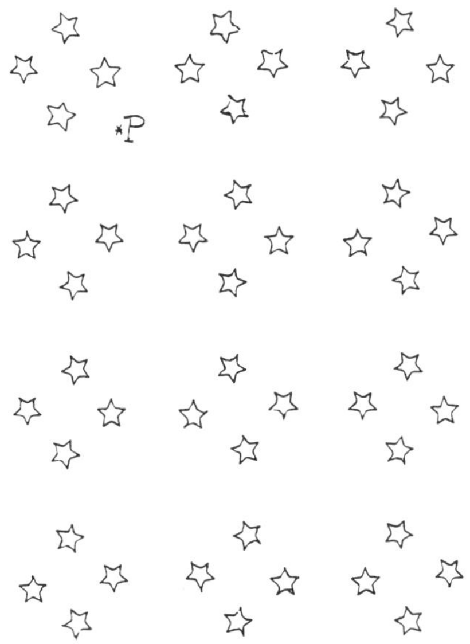

*P

Figure 12-1 ***Periodically repeating pattern***

$33'33''$ and $22'2''2'''$. Each of the symbols $2, 2', 2'', 2''', 3, 3', 3'', 4, 4'$ and 6 denotes a separate rotocomplex whose points are, of course, symmetrically related to each other. Enantiomorphically paired rotocenters are considered as members of the same rotocomplex: For example, 4 and $\tilde{4}$ belong to the same rotocomplex, but 4 and $4'$ belong to *distinct* rotocomplexes.

A *mesh* is a triangle or (in the case of $22'2''2'''$) a parallelogram whose vertices are all either distinct or enantiomorphs, and which contain no rotocenter internally or on a boundary, other than at a vertex. It is difficult to define a mesh for the case $1\infty\infty$, as all rotocenters are infinitely far away. The mesh for $22'\infty$ is a zone bounded by two parallel lines and a line intersecting those parallel lines at two distinct two-fold rotocomplexes. For 236 the mesh is a right triangle having angles 30° and 60°. The $244'$ mesh is an isosceles right triangle, while that for the $33'3''$ system is an equilateral triangle. The mesh for $22'2''2'''$ is a parallelogram. Since every point in the plane is associated with one mesh or another, meshes tessellate the plane.

Q: *ON A SECOND PIECE OF TRACING PAPER DRAW THE MESHES FOR FIGURE 12-1. MARK THE ROTOCOMPLEXES. CALL THIS FIGURE 12-1 (b). SUPERIMPOSE THE TWO PIECES OF TRACING PAPER (FIGURES 12-1 (a) AND 12-1 (b)) TO COMPARE THE ARRAYS OF POINTS AND THE REGIONS FOR THIS PATTERN.*

In Chapter VIII we noted that when all rotocenters lie on mirrors, these mirrors are mesh boundaries, whereas when rotocenters are enantiomorphically paired, meshes will be bisected by mirror lines. Unless the mesh boundaries are mirror lines, they do not necessarily need to be straight lines, but there appears to be little advantage to choosing curved-edged meshes.

A *lattice* is the array of all points related by *translational* symmetry. Within a lattice *complex*, which constitutes all points related by translational, rotational or reflection symmetry, a lattice is a subset: A lattice complex is generally composed of several lattices, as membership in a lattice is more restrictive (translational symmetry *only*) than membership in a lattice complex (translational, rotational or reflection symmetry).

Q: *PLACE FIGURE 12-1 (a) BACK OVER FIGURE 12-1. PUT A CIRCLE AROUND THOSE ASTERISKS WHICH DENOTE POINTS BELONGING TO THE SAME LATTICE AS DOES THE POINT P. NOW SLIDE FIGURE 12-1 (a) PARALLEL TO ITSELF, AND STOP WHEN THE POINT P IS AGAIN COVERED BY AN ASTERISK. DO ALL POINTS NOW COVERED BY AN ASTERISK HAVE THE SAME CONTEXT? WHAT ABOUT THE POINTS COVERED BY A CIRCLED ASTERISK?*

Every point in a periodically repeating pattern belongs to a lattice; there are therefore actually infinitely many lattices. However, you will have noted by sliding Figure 12-1 (a) over Figure 12-1 that the lattice to which a given point belongs is actually congruent to the lattice to which every other point belongs. We therefore speak of *the* lattice of a periodically repeating lattice when we actually mean the representative lattice characteristic of the translational symmetry of this pattern. Note that there is no such characteristic lattice *complex* for periodic structures: The shape of a lattice complex depends on the location of its points relative to the rotocenters.

Associated with the concept of a lattice is that of the unit cell. A *unit cell* is a polygon which tiles the plane by *translation*, and which contains at least one point of every lattice. A unit cell does not necessarily need to be bounded by straight edges. Because it tiles the plane by translation, it must have an even number of edges, and opposite edges must be congruent to each other. A triangle, for instance, could not be a unit cell.

Q: *TAKE A THIRD PIECE OF TRACING PAPER, AND TRACE ON IT A UNIT CELL WHICH CONTAINS THE POINT P BUT NO OTHER POINT OF THE SAME LATTICE. CALL THIS DRAWING FIGURE 12-1 (c). COULD THE DIRICHLET DOMAINS OF FIGURE 12-1 (a) BE UNIT CELLS?*

In Chapter XI we found triangular Dirichlet domains, proving that not every Dirichlet domain may be a unit cell. The complete array of points which has such triangular Dirichlet domains is not a lattice: The contexts of these points are not always in the same orientation. Dirichlet domains of a *lattice* are necessarily unit cells. The definition of a unit cell does not exclude the possibility that more than one point of the same lattice lies inside the same unit cell. When this happens, we call the unit cell a *multiple* unit cell; when only a single lattice point lies inside a unit cell, the cell is called *primitive*. In Figure 12-1 (c) you should have drawn a primitive unit cell. A Dirichlet domain of a *lattice* is necessarily a primitive unit cell.

In Figure 12-2 we have drawn a square straight-edged unit cell for the pattern in Figure 12-1. Such a unit cell is useful for crystallographers. It is multiple, for there are several points equivalent to P in this unit cell: One is at the center, and four are at the corners. Since the points at the four corners are shared by four unit cells, each of these counts as one quarter. As a result, the four points count as a single lattice point. There are therefore two lattice points in the unit cell of Figure 12-2: This is a double unit cell.

Q: *DRAW A PRIMITIVE UNIT CELL FOR THE PATTERN OF FIGURE 12-1.*[2] *WOULD A PARALLELOGRAM OR EVEN A RHOMBUS (EQUILATERAL PARALLELOGRAM) WORK AS A UNIT CELL IN THIS EXAMPLE?*

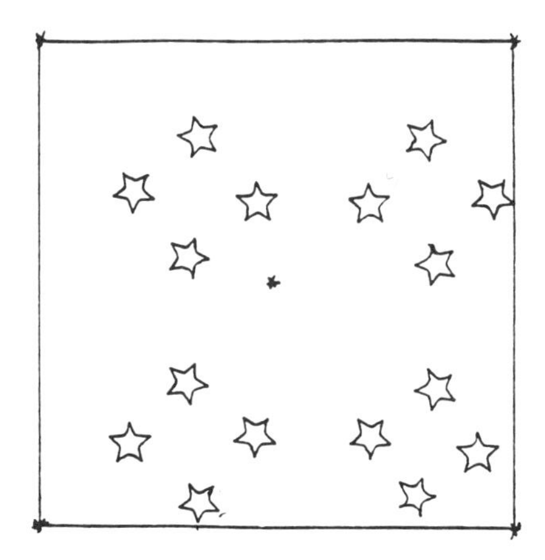

Figure 12-2 ***A square unit cell for Figure 12-1***

The concept "unit cell" is useful not only to crystallographers, but also to designers who print a pattern onto paper or textile from a roller; such patterns necessarily have translational symmetry. The information for the entire pattern is contained in a single unit cell which is repeated by translation. A multiple unit cell contains this information several times: In a multiple unit cell there are several points which are related to each other by translational symmetry.

Finally we shall consider here a *fundamental region*, a region which contains all the information regarding a periodically repeating pattern just once. A Dirichlet domain is one example of such a region, as is a primitive unit cell. (Remember that a Dirichlet domain of the *lattice* is a primitive unit cell.) A *mesh* may or may not be a fundamental region. We saw previously (Chapter VI) that the number of meshes meeting at a rotocenter equals exactly twice the symmetry value of that rotocenter. When all rotocenters lie on mirror lines, adjoining meshes are enantiomorphically related; in this case a single mesh contains all the information which may be repeated symmetrically (by rotation, translation or reflection) to generate the complete pattern. When rotocenters are enantiomorphs, meshes are bisected by mirror lines, and fundamental regions do not coincide with meshes.

Q: *PLACE A PIECE OF TRACING PAPER OVER FIGURE 12-1, AND TRACE A MESH AND A FUNDAMENTAL REGION.*

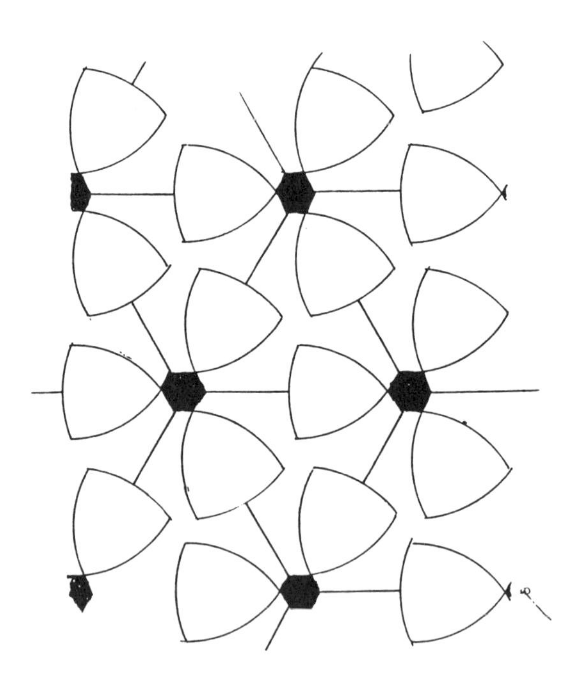

Figure 12-3 ***A periodically repeating pattern***

In the absence of enantiomorphy any pair of adjoining meshes may be combined to form a fundamental region, as such meshes may now contain separate independent information.

Q: *PLACE A PIECE OF TRACING PAPER OVER FIGURE 12-3, AND DRAW:*

- *A. A LATTICE;*
- *B. A LATTICE COMPLEX;*
- *C. ALL ROTOCENTERS;*
- *D. A MESH;*
- *E. A FUNDAMENTAL REGION;*
- *F. A RECTANGULAR UNIT CELL;*
- *G. A PRIMITIVE UNIT CELL;*
- *H. A DIRICHLET DOMAIN OF THE LATTICE.*

Q: *MEASURE THE AREA OF THE TILE IN FIGURE 10-4.*[3]

NOTES

[1] Figures 12-4 (a) and 12-4 (b) show some points equivalent to P, together with their Dirichlet Domains. They also show a mesh; the symmetry of the pattern in Figure 12-1 is $\underline{2}4\tilde{4}$.

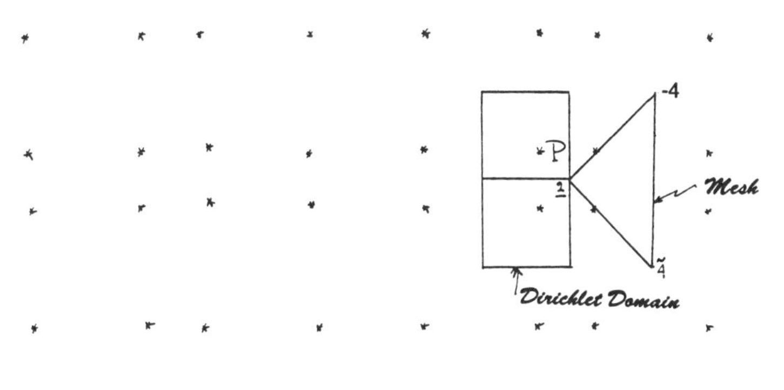

Figure 12-4 (a) Figure 12-4(b)

[2] A diagonally oriented square whose vertices are at the midpoints of the edges of the double unit cell of Figure 12-2.

[3] Note that the tile of Figure 10-4 may be transformed into a straight-edged tile by scooping out one portion judiciously and transferring it elsewhere with the result that a simpler tile in the form of a parallelogram or rectangle is formed whose area equals that of the original tile but is more easily measured. Refer to Chapter II, however, for the choice of a unit area.

XIII

A Look at Infinity

Four bugs are located at the four corners of a square. Each looks at a bug nearest to it in a clockwise direction. Each moves toward that neighbor, all four bugs moving at the same speed at any given moment, although that speed does not necessarily remain constant in time.

Q: *WILL THE BUGS EVER REACH EACH OTHER, AND IF SO, HOW FAR WILL THEY NEED TO TRAVEL BEFORE THEY REACH EACH OTHER?*

Initially, each bug will move along the edge of the square toward its neighbor. Very soon it will discover that its target has moved, and that it will need to change direction. Let us arbitrarily call the edge length of the original square unity, and let us assume that each bug has traveled a tiny distance δs, before discovering that it has to modify its direction of travel. At that time the bugs will again find themselves at the corner of a square, but a slightly smaller one, (dotted lines in Figure 13-1), which is also

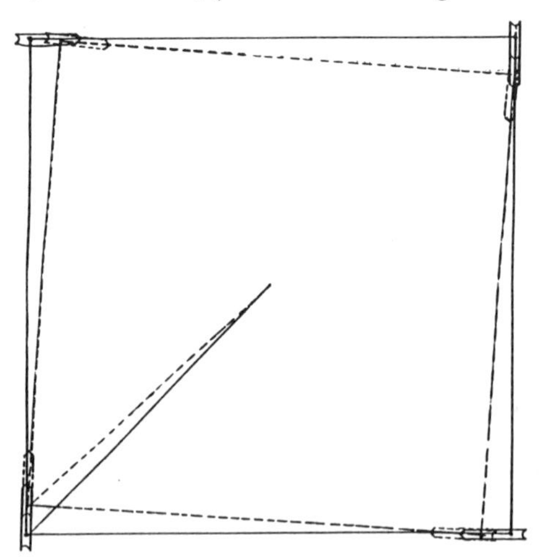

Figure 13-1 ***Bugs***

slightly tilted with respect to the original (square solid lines in Figure 13-1). Once more, the bugs will need to adjust their direction in order to keep tracking their targets. At any instant in the process the situation is the same: The bugs are at the corners of a square, which shrinks as well as rotates as the bugs continue tracking each other. Figures 13-2 (a) and (b) show us one of the bugs somewhere on its way.

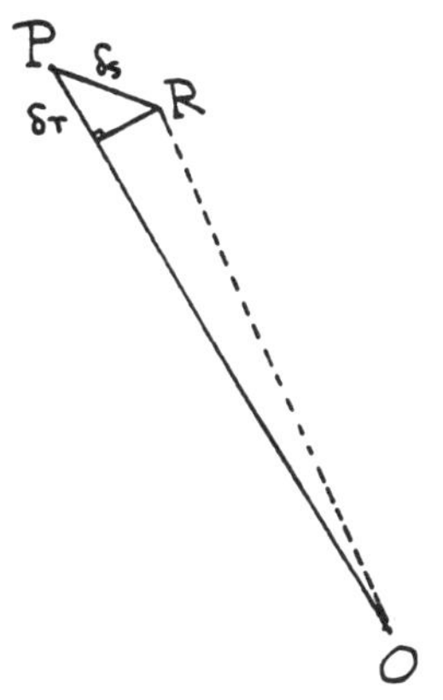

Figure 13-2 (a) ***A bug on its way***

Q: *WHAT IS THE ANGLE BETWEEN THE DIRECTION OF TRAVEL OF A BUG AND THE LINE JOINING ITS POSITION TO THE CENTER OF THE SQUARE?*

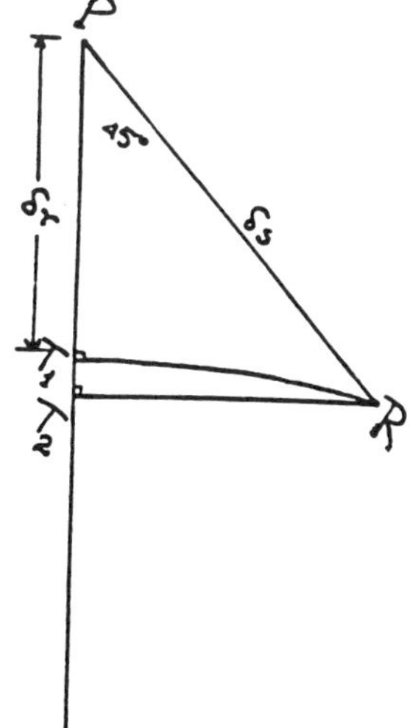

Figure 13-2 (b) ***Magnification of a portion of Figure 13-2 (a)***

It will be convenient to describe the position of the bug in terms of the location of the center of the original square, O, which is also the center around which the square rotates. The initial distance of each bug from this center is $\frac{1}{2}\sqrt{2}$; at any subsequent time we shall label its distance from the center r; we call it the *radial coordinate*. At any moment the direction of travel is along the edge of the square of that moment, that is to say, at 45° to the *radial direction*, the direction from the bug to the center of the square. As drawn in Figure 13-2 (a), by traveling the distance δs from point P to point R, the bug will have come closer to the center of its square by a distance equal to the difference between the lengths of the lines OP and OR, δr. It will also have caused its square to rotate by a small angle $\delta\theta$.

In Figure 13-2 (b), a magnification of the triangle around points P and R in Figure 13-2 (a), we have drawn a circular arc RT_1 whose center is at O, whose radius is OR and which meets the line OP at T_1. The distance PT_1 then equals δr. We have also drawn a line RT_2 perpendicular to OP, T_2 lying on the line OP. The distance between the points T_1 and T_2 is very small indeed. In the next chapter we shall return to it in a more quantitative manner; for the present simply note that the distance T_1T_2 is small even compared to the already tiny distances δs and δr. The magnitude of the angle $\delta\theta$ is determined by the rapidity with which the bugs are able to readjust their direction of travel. If the bugs' reaction is sufficiently fast, then we might even neglect the distance T_1T_2 altogether; in that case we may approximate the difference between the lengths of OP and OR by the length PT_2 rather than PT_1.

If the reaction speed of the bugs is sufficiently high, then we shall perceive their path as a smooth spiral. There is a correlation between our ability to detect discrete segments along the spiral and the reaction speed of the bugs: We can say that within our power of resolution the difference between the lengths of OP and OR may be approximated by δr as long as the bugs' reaction speed is sufficiently high.

The triangle PRT_2 is right isosceles; as a result, the distances δs and δr are, within our level of resolution, related by the equation

$$\delta s = \sqrt{2}\,\delta r \qquad (13-1)$$

Let us now suppose that we were photographing the progress of the bugs with a camera which rotates with the same angular velocity as does the square, and takes a picture every time it rotates through an angle $\delta\theta$. From the camera's eye the bugs would appear to be crawling straight toward the center; between successive camera-shots each bug would come a distance δr closer to the center O, successive quantities δr becoming smaller (Figure 13-3). We have just seen, however, that for each interval between camera shots the actual distance δs traveled equals $\sqrt{2}\delta r$.

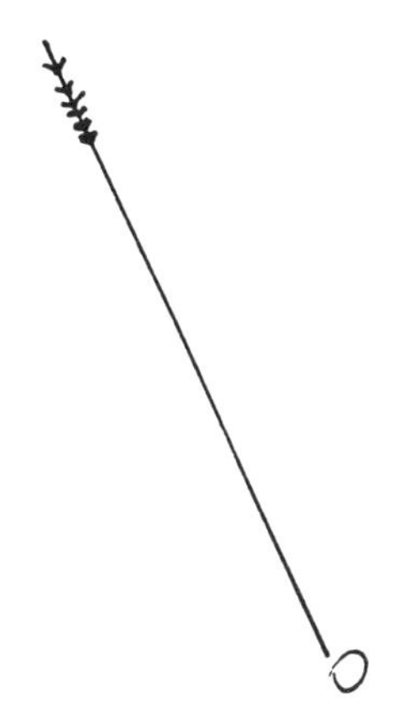

Figure 13-3 ***A bug's progress observed from a rotating frame of reference***

Q: *ADDING ALL THESE BITS OF DISTANCE TOGETHER, WHAT WOULD THE TOTAL PATH LENGTH BE UNTIL THE BUGS WERE TO MEET?*

Ultimately the bugs would hope to meet at the center, O. The path they will follow consist of a succession of tiny line segments, each being exactly $\sqrt{2}$ times the corresponding decrease in the radial coordinate, δr. The sum of these segments would therefore amount to $\sqrt{2}$ times the initial distance from the center, that is to say $\sqrt{2} \times \frac{1}{2}\sqrt{2}$, which is unity. In other words, the total distance covered by the bugs spiralling inward toward each other is exactly the distance which originally separated them!

Q: *IF THE BUGS ARE TO TRAVEL AT A CONSTANT LINEAR SPEED, THEN WHAT WILL BE THEIR ANGULAR VELOCITY AROUND THE CENTER OF THE SQUARE BE WHEN THEY APPROACH THAT CENTER?*

Before rejoicing at the easy path toward the bugs' reunion, let us take a closer look at what happens when the bugs approach each other. Presumably, they will meet shoulder to shoulder before reaching the center O, but the smaller they are, the more times they need to rotate around the center. In order to maintain a reasonable linear speed, they must rotate at increasing angular velocity, because their linear speed equals at any time the product of the radial coordinate r with the angular velocity. The smaller the bugs, the faster they need to rotate, and ultimately they may fly apart because of the enormous "centrifugal force" they experience.

Obviously, they will need to decrease their linear speed to survive, but if they slow down too much, they will never quite reach each other.

Buckminster Fuller used to complain about being taught that a point has no dimension, yet a lot of points placed next to each other will constitute a line, which in turn has no width! He said that if he drew a line, and looked at it through a magnifying glass he could see its finite width; moreover, sufficient magnification would reveal that the line had a structure, being made up of graphite particles and x-ray examination of these particles would discover the disposition of carbon atoms in the graphite. Fuller maintained that there is no such thing as a single dimensionless point, since everything has structure. He would hasten to declare that our bugs, being finite in size, will ultimately meet shoulder to shoulder, and that our concern about their meeting at the center is only due to our wanting to make them into dimensionless points. Fuller also stated that there is no such thing as a circle, because what we call a circle is actually a polygon having enormously many sides. By the same token he would point out that the spiral traced by our bugs consists of many small line segments, each being of a finite length determined by the reaction speed of the bugs.

Fuller's discomfort with the infinite and infinitesimal is part of a tradition stretching back to Greek antiquity and continuing into modern cosmology: At one Harvard University's 350th anniversary symposia the question was raised concerning our expanding universe: "If it is finite, then what does it expand into?" If we think of the circle as the locus of all points equidistant from a given point, then the number of points that will obey that condition is certainly not finite! But we can only use that definition as long as we believe in a dimensionless point.

In Chapter VI we encountered infinitefold rotational symmetry, which we interpreted as translational symmetry. Here we used Euclid's supposition that parallel lines meet at infinity, a postulate which has been called into question by modern (non-euclidean) mathematicians.

Structure is hierarchical. The structures with which Fuller was dealing are physical structures: We find with increasing power of resolution that what had appeared to be indivisible ("atom" means that which cannot be cut!) has a structure of its own. The smaller our "indivisible" particle is, the more energy appears to be needed to observe its internal structure. Finding ourselves running out of resources to supply the energy needed to split our present "elementary" particles, we use solar and astral laboratories to explore elementary structure further. However, the question of whether we shall ever find a truly elementary particle may well be moot, because we may need all the energy in the universe to split it in order to determine its structure.

We are fortunate in having limited resolving power. Imagine being endowed with x-ray vision: Instead of seeing the illustrations in this book as line drawings, we would see a mass of carbon and other particles. Limited resolving power enables us to observe structure hierarchically: We can enjoy architectural beauty without needing to know the atomic configuration of the granite used. Limited resolving power allows us to experience the path traced by the bugs as a smooth spiral, albeit granting that a finite reaction time on the part of the bugs actually causes it to be a sequence of line segments.

Limiting resolving power also allows us to make approximations such as neglecting the distance T_1T_2. Mathematicians would say something like: "The ratio of the distance T_1T_2 to δs goes to zero as $\delta\theta$ goes to zero." Buckminster Fuller would object that none of them can go to zero and that no wonder mathematicians run into indeterminacies by taking ratios of quantities which they suppose to become zero. We will not do violence to either view if we realize that when a mathematician lets a quantity go to zero (s)he actually means that within the pre-set power of resolution the quantity appears to have vanished. In the case of the bugs' spiral we can choose $\delta\theta$ sufficiently small that the distinction between the points T_1 and T_2 can no longer be made within our level of resolution.

Structure is hierarchical. Physical structures, we have noted, are hierarchical because of the different levels of resolving power of the apparatus or the wavelength of the radiation used to observe them. Mathematical structures also are hierarchical: We noted that in Figure 13-2 the lengths δs and δr are tiny, but that compared to them the distance T_1T_2 is much tinier yet, hence negligible within our level of resolution. This hierarchy is behind the power of calculus: We shall see in further examples in future chapters that the relative smallness of quantities allows us to make simplifications, and also leads us to the so-called irrational numbers with which Buckminster Fuller was so uncomfortable.

Notes

[1] Edmondson, Amy C.: *A Fuller Explanation*, Chapter 2: "The Irrationality of Pi," pp. 15–25. Design Science Collection, A. L. Loeb, series ed., Birkhäuser, Boston (1987).

XIV

An Irrational Number

The bugs studied in the previous chapter generated a curve which makes a constant angle, namely 45° with the direction toward the origin (the radial direction). We could have used six bugs at the corners of a regular hexagon, in which case they would have travelled at 60° to the radial direction.

Q: *HOW FAR WOULD THE BUGS AT THE CORNER OF A REGULAR HEXAGON HAVE TO TRAVEL IN ORDER TO MEET?*

As we shall see presently, spirals making a constant, not necessarily 45°, angle with the radial direction, are very common and important; for this reason we shall extend our discussion to this more general application.

Figure 14-1 shows the radial and tangential components of a spiral making an angle α with the radial direction. The radial component PT_2 and the tangential component δu are related to the angle α by equation 13-1:

$$\delta u \;=\; \tan\alpha \cdot PT_2 \qquad (14-1)$$

If we travel from point P to point R, then the radial coordinate decreases from OP to OR. In the previous chapter we assumed that the tangential component is, within our level of resolution, equivalent to a circular arc, and indicated that we would more closely look at that assumption. Figure 14-1 shows that a circular arc would land us at point T_1 rather than at T_2. The respective distances of these two points from the origin are

$$OT_1 \;=\; r \;-\; \delta r \qquad (14-1a)$$

$$OT_2 \;=\; (r \;-\; \delta r)\,\cos\delta\theta\,. \qquad (14-1b)$$

The factor $\cos\delta\theta$ very rapidly approaches unity for small values of $\delta\theta$. For instance, for an angle $\delta\theta$ of 0.01 radians (about three-fifths of a degree), this factor is 0.99995; for an angle one-tenth of this, the factor is 0.9999995. Every decrease of angle $\delta\theta$ by a factor of ten results in two additional nines between the decimal point and the digit 5. Clearly, then, we may safely

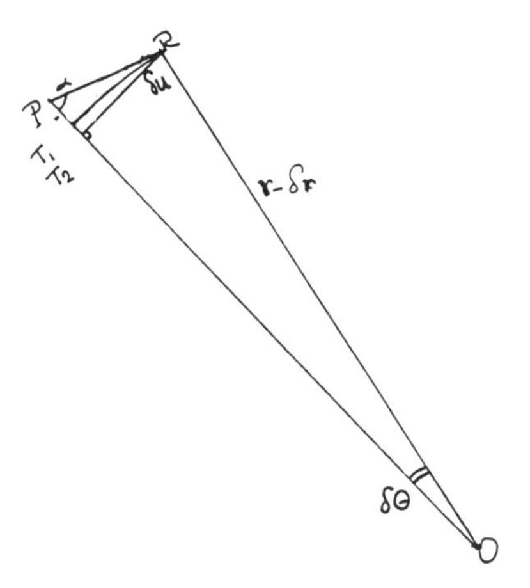

Figure 14-1 ***Radial and tangential components of spiral making a constant angle with the radial direction***

assume that if the spiral looks smooth within our level of resolution, the difference between OT_1 and OT_2 may be disregarded, and we can make the approximation that a small change in angle $\delta\theta$ results in a decrease of the radial distance by the distance PT_2. We may similarly approximate the component δu by the arc length $(r - \delta r)\delta\theta$. Therefore we find, with the aid of Equation 14-1,

$$\delta u = (r - \delta r)\delta\theta = -\delta r \tan\alpha .$$

Note that the minus sign on the right hand of this equation is due to the fact that in Figure 14-1 the radius r *decreases* as the angle θ *increases*. For a smooth spiral, δr may be neglected compared to r, so that

$$\frac{1}{r}\frac{\delta r}{\delta\theta} = -\tan\alpha . \qquad (14-2)$$

The ratio of δr to $\delta\theta$ indicates the rate with which the radial distance r changes as the angle θ changes. When this ratio is divided by r, the result is the fractional or *relative* rate of change of r with θ. Equation 14-2 tells us that for a spiral which makes a constant angle with the radius, the relative rate of decrease of the radius with the angle is also constant, and equal to the tangent of the angle which the spiral makes with the radius.

In the previous chapter we considered a special spiral which makes an angle of 45° with the radius, in connection with the friendly bugs. We found that the bugs need only travel a finite distance to reach the origin, a surprising result. However, we are not yet convinced that the bugs' path has world shaking consequences. If we can assess the importance of processes whose relative change rate is constant, then we shall also know how powerful the "constant angle" spiral is as a graphic representation. Before returning to geometrical representations, we shall encounter some properties of numbers, and a remarkable constant which cannot be expressed as a ratio of integers.

Let us suppose that r_0 represents an amount of capital which is invested at a time t_0 at an interest rate of $a\%$ per year. Then after one year the capital will be given by r_1:

$$r_1 = r_0(1 + 0.01a)\,.$$

If the capital r_1 is in turn reinvested at the same interest rate, then after a second year the capital will have attained a value r_2:

$$r_2 = r_0(1+0.01a)^2\,.$$

Q: *CALCULATE YOUR CAPITAL AFTER TWO YEARS IF YOU STARTED WITH $100 AT 10% INTEREST: a) IF YOU KEEP THE INTEREST AFTER A YEAR, AND b) IF YOU ADDED THE INTEREST AFTER YEAR TO THE CAPITAL.*

The process of reinvesting the accrued capital is called *compounding*. The example just given compounds interest annually. One could, however, also compound semi-annually. After a half-year, the capital would have been $r_{\frac{1}{2}}$:

$$r_{\frac{1}{2}} = r_0(1 + 0.005a)\,.$$

Then, if the resulting capital is reinvested for another half-year, it will at the end of a year have grown to:

$$r_1 = r_0(1 + 0.005a)^2\,.$$

To compare the results of annual and semi-annual compounding, let us assume an annual rate of interest of 10% and an initial investment of $100. Annual compounding produces after one year $r_1 =\$100(1+0.1) =\$110$. Semiannual compounding produces after one year $r_1 =\$100(1+0.05)^2 =$ $110.25. Since semiannual compounding yields an extra quarter, what would quarterly compounding yield us? The answer is $\$100(1+0.025)^4 =$ $110.38. Monthly compounding: $110.47. Note that the greatest profit resulted when we changed from annual to semi-annual compounding; it would appear as if daily or even hourly compounding would not have as radical a result.

With modern computer facilities available, it would be possible for a bank to compound interest at very small time intervals, say δt, expressed as a fraction of a year. After a year initial capital r_0 will have grown to

$$r_1 = r_0(1 + 0.01a\,\delta t)^{\frac{1}{\delta t}}\,. \qquad (14-3)$$

This expression behaves curiously for very short time intervals δt. The expression in parentheses will be a little bit more than unity, whereas the exponent will be enormously large. We know on the one hand that unity raised to any power, no matter how large, will always yield unity, whereas any number larger than unity will yield a very large number when raised to a larger power. This knowledge does not help us in evaluating the expression on the right hand side of equation 14-3, for even though the expression in parentheses may be very close to unity, it is larger than unity, and hence might increase substantially when raised to a large power. Just as in the case of the bug trajectory, we are dealing with a hierarchy of magnitudes: We use expressions *large, very large*, and *substantial*, but the meaning of the latter is as yet *indeterminate*. To resolve this indeterminacy, let us put equation 14-3 in a slightly different form by scaling the time variable as follows:

$$\tau = 0.01\,at, \quad \delta\tau = 0.01a\,\delta t\,.$$

Equation 14-3 then becomes

$$r_1 = r_0[(1 + \delta\tau)^{\frac{1}{\delta\tau}}]^{0.01a}\,. \qquad (14-4)$$

In this form we can evaluate the expression within the brackets for smaller and smaller time intervals; the exponent outside the brackets will, of course, depend on the particular interest rate, but will not change with the time interval chosen. We need to raise $(1 + \delta\tau)$ to a very high power; to do so, we use a generalized expression for "power raising." Let us first write out some low powers of $(1 + \delta\tau)$:

$$\begin{aligned}
(1 + \delta\tau)^2 &= 1 + 2\delta\tau + (\delta\tau)^2, \\
(1 + \delta\tau)^3 &= 1 + 3\delta\tau + (\delta\tau)^2 + (\delta\tau)^3, \\
(1 + \delta\tau)^4 &= 1 + 4\delta\tau + 6(\delta\tau)^2 + 4(\delta\tau)^3 + (\delta\tau)^4, \\
(1 + \delta\tau)^5 &= 1 + 5\delta\tau + 10(\delta\tau)^2 + 10(\delta\tau)^3 + 5(\delta\tau)^4 + (\delta\tau)^5\,.
\end{aligned}$$

Q: *DO YOU SEE ANY CONNECTION BETWEEN THE EXPONENT ON THE LEFT HAND SIDE AND THE NUMBER OF TERMS ON THE RIGHT?*

From these expressions we can infer some generalizations. In the first place, the number of terms on the right hand side equals the exponent on the left side plus one. Secondly, on the right hand side the first term always equals unity, while the coefficient of the second term equals the exponent on the left side. The coefficients of succeeding terms are not so conveniently generalized, but they are generated with the aid of Pascal's triangle.

Table 14-1: ***Pascal's Triangle.***

1 1
1 2 1
1 3 3 1
1 4 6 4 1
⋮

Pascal's triangle is shown in Table 4-1; in it each number equals the sum of the nearest two numbers above it. The lth term in the nth row is given by the expression

$$\frac{n!}{(n-l+1)!(l-1)!}$$

where the notation $n!$, called *n-factorial*, denotes the product of all integers from 1 through n, and $0! = 1$. For instance, for $n = 5, l = 3$ the corresponding number in Pascal's triangle is:

$$\frac{5 \cdot 4 \cdot 3 \cdot 2 \cdot 1}{(3 \cdot 2 \cdot 1)(2 \cdot 1)} = 10\,.$$

Comparison with the expansion of powers of $1 + \delta\tau$ illustrates that the nth row in Pascal's triangle lists just the successive coefficients for the nth power of $1 + \delta\tau$, so that we may write

$$\begin{aligned}(1+\delta\tau)^n &= 1 + n\,\delta\tau + \frac{1}{2}n(n-1)(\delta\tau)^2 \\ &+ \frac{1}{6}n(n-1)(n-2)(\delta\tau)^3 + \ldots \\ &+ \frac{n!}{(n-l+1)!(l-1)!}(\delta\tau)^{l-1} + \ldots + (\delta\tau)^n\,.\end{aligned}$$

Actually, in our example, we must raise $1 + \delta\tau$ to the power $1/\delta\tau$, hence here $n = 1/\delta\tau$. Note that the second term in the expansion, $n\delta\tau$, then will always be unity. Hence:

$$\begin{aligned}(1+\delta\tau)^{1/\delta\tau} &= 2 + \frac{1}{2}(1-\delta\tau) + \frac{1}{6}(1-\delta\tau)(1-2\delta\tau) + \ldots \qquad (14-5) \\ &+ \frac{1}{l!}[(1-\delta\tau)(1-2\delta\tau)\ldots\{1-(l-1)\delta\tau\}] + \ldots + (\delta\tau)^{\frac{1}{\delta\tau}}\,.\end{aligned}$$

This equation gives us a great deal of useful information. Since the terms l and $n\delta\tau$ have merged to give 2, the number of terms on the right side now equals the exponent in the left hand side; therefore, when $\delta\tau$ is very small, hence $1/\delta\tau$ very large, there will be a very large number of terms on the right hand side. Moreover, successive terms will decrease rapidly in magnitude as a result of the $l!$ in the denominator, the last term being the very small fraction $\delta\tau$ being raised to an enormous power, hence very tiny indeed. Furthermore, the numerator in each term consists of the product of factors each of which is only slightly less than unity.

In Table 14-2 we list values of $1/l!$, exact to seven decimal places, the level of resolution which we desire to establish.

Table 14-2: ***Reciprocal Factorials.***

l	$1/l!$
2	0.5000000
3	0.1666666
4	0.0416666
5	0.0083333
6	0.0013888
7	0.0001984
8	0.0000248
9	0.0000027
10	0.0000003

If we wish to determine $(1+\delta\tau)^{1/\delta\tau}$ to only seven decimal places, then additional terms will not be necessary.[1] Moreover, we may choose $\delta\tau$ sufficiently small that the factors $(1-\delta\tau)$, $(1-2\delta\tau)$,... will differ from 1 by a quantity sufficiently small that they will not contribute to the first seven decimal places. In that case, we may say that within our pre-set level of resolution the expression $(1+\delta\tau)^{1/\delta\tau}$ equals the sum of values in the second column of Table 14-2 added to 2, that is to say 2.7182815.

We could, of course, refine our level of resolution further, say to one million decimal places. This would necessitate a much smaller value of $\delta\tau$, hence many more terms in the summation. The fact is that increased refinement would require ever-increasing numbers of decimal places to express $(1+\delta\tau)^{1/\delta\tau}$ in decimal notation. The number 2.7182815 would acquire longer and longer trains of digits; its magnitude would not change appreciably, however. As we shall see presently, a number which has infinitely many digits behind the decimal point in a non-repeating pattern is necessarily irrational. (A *ratio*nal number is one which is expressible as a *ratio* of integers, an irrational one is not). The expression $(1+\delta\tau)^{1/\delta\tau}$ is rational if $\delta\tau$ is. Therefore if we continue to decrease

$\delta\tau$, we continue to approach, but will never reach, an irrational number which to seven decimals may be approximated by 2.7182815. This important number is designated by the symbol e.

We are confronted with the relationship between the irrationality of numbers and the level of resolution, and we have an illustration of why Buckminster Fuller, who insisted on the discreteness of nature, was uncomfortable with irrational numbers. A number having a finite number of digits behind the decimal point is necessarily rational, as it can be expressed as a ratio of an integer and a power of ten. The converse is not true: Rational numbers such as 1/3 and 5/12 have infinitely many digits behind the decimal point when expressed in decimal notation. Such rational numbers, however, differ from irrational numbers in an important respect: Rational numbers in decimal notation will inevitably, after a finite number of decimal places, repeat the same sequence of digits indefinitely. For example,

$$\begin{aligned} 1/2 &= 0.500000000\ldots \quad && (0 \text{ repeats indefinitely}), \\ 1/3 &= 0.333333333\ldots \quad && (3 \text{ repeats indefinitely}), \\ 5/12 &= 0.416666666\ldots \quad && (6 \text{ repeats indefinitely}), \\ 473/333 &= 1.420420420\ldots \quad && (\text{The sequence 420 repeats indefinitely}). \end{aligned}$$

A repeating decimal representation is inevitably rational, and decimal representation of an irrational number will have infinitely many digits behind the decimal point, but never an indefinitely repeating sequence. This generalization will be understood when we look at the examples just given. As illustrated in those examples, a repeating sequence of digits behind a decimal point may be put in the generalized form.

$$10^{-k}[a \times 10^{-m} + a \times 10^{-2m} + \ldots + a \times 10^{-nm} + \ldots].$$

Here a represents the sequence of repeating digits, k is the number of digits behind the decimal point before the pattern begins to repeat, and m is the number of digits in the repeating pattern. For the four examples given above, the decimal representation of 1/2 has $k = 1, m = 1, a = 0$, the decimal representation of 1/3 has $k = 0, m = 1, a = 3$, the decimal representation of 5/12 has $k = 2, m = 1, a = 6$, and the decimal representation of 473/333 has $k = 0, m = 3, a = 420$. This generalized form may be rewritten as follows:

$$a \times 10^{-(k+m)} S, \quad \text{where} \quad S \equiv 1 + 10^{-m} + 10^{-2m} + \ldots. \qquad (14-6)$$

Multiply both sides of the latter identity by 10^{-m}:

$$10^{-m} S \equiv 10^{-m} + 10^{-2m} + \ldots. \qquad (14-7)$$

Subtracting equation (14-7) from (14-6), we obtain $(1 - 10^{-m})S = 1$, hence the generalized form of repeating decimal representation is

$$a/(10^{k+m} - 10^{k}).$$

This expression is necessarily rational, so that we have demonstrated that repeating decimal representations (including, of course, representations having a finite number of digits behind the decimal point) are necessarily rational.

Now let us return to equation 14-3, which expresses the amount of capital after a year of compounding interest at time intervals of δt. As we shorten this time interval, we will eventually come as close as we wish to

$$r_1 = r_0 e^{0.01a} . \qquad (14-8)$$

This expression gives an upper limit to what one may expect to achieve by shortening the compounding interval indefinitely.

Let us follow the compounding process from the beginning. An amount of capital r_0 will, after a short time interval δt at interest of $a\%$ per year, accrue to $r_{\delta t} = r_0(1 + 0.01a\delta t)$, where δt is expressed as a fraction of a year. The increase in capital, δr, is $0.01 r_0 \, a\delta t$, hence:

$$\frac{1}{r_0}\frac{\delta r}{\delta t} = 0.01\, a . \qquad (14-9)$$

At any moment of compounding interest the *relative* rate of increase is constant, and determined solely by the annual interest rate. This observation implies that the curve described by the bugs may indeed have more general applicability, and could, for instance, be a suitable graphic representation of the increase in capital as a result of frequent compounding of interest. Note, however, that the radius for the bugs decreased with increasing angle θ, whereas capital *increases* with time as a result of compounding interest. This means that the spiral graph for interest compounding will have an angle α greater than ninety degrees, or that we must use the bugs' graph backwards, i.e., with time represented in the negative θ direction.

If we continue compounding interest for a time period t equal to an integral multiple of δt, then the capital will be given by

$$r(t) = r_0(1 + 0.01a\delta t)^{t/\delta t} .$$

Scaling once more, and letting the compounding intervals decrease, we approach

$$r(t) = r_0 e^{0.01at} . \qquad (14-10)$$

Equation (14-10) defines an important function, the *exponential function.* It represents processes whose relative growth rate is constant, for instance, the decrease of the distance from the bugs to the origin, and the growth of capital on compounded interest. The exponential function is smooth in that it represents compounding of interest at such small time intervals that within our level of resolution the capital appears to increase smoothly; analogously it represents bugs which react so quickly to the change in position of their targets that their paths within our level of resolution appears smooth.

In this chapter we have introduced the principles of the mathematics of continuous processes and structures, which is known as *calculus*. We have observed that a belief in continuous structures corresponds to a faith in the existence of idealized functions and irrational numbers which can only be approached arbitrarily closely, but never reached. In the next chapter we shall bring the use of terms such as "within our level of resolution" into our conformance with the nomenclature of calculus, and we shall consider some further applications of these principles to growth processes. Many students in design science appear to be uncomfortable with calculus. For this reason we introduce this material with the aid of specific graphic examples. The reader familiar with calculus may skip these chapters or skim over them, as there may be some images helpful in teaching this branch of mathematics.

Notes

[1] Strictly speaking, we need mathematicians to prove that the many tiny terms following the tenth one might not add up to make a substantial contribution. Such a proof is outside the scope of the present volume, but the reader is assured that in this instance the terms diminish so rapidly in magnitude that they will not contribute to the first seven decimal places.

XV

In Chapters XIII and XIV we dealt with issues of discrete and continuous structures, rational and irrational numbers, and recognized the relationships between them. These are actually the fundamental concerns of calculus; if they are understood, then the remainder of calculus is essentially a question of notation.

Basic to calculus is the notion of *level of resolution*. We came to realize that within a given pre-set level of resolution quantities can be organized in a hierarchy of smallness, so that one can discard some because they are too small to be observable within that pre-set level of resolution. In particular, we defined an irrational number e which can be approximated to any desired accuracy by the summation

$$e = 1 + \frac{1}{2} + \frac{1}{6} + \ldots + \frac{1}{l!} + \ldots$$

The three dots at the end of this last line mean that one can add terms to the summation until the sum no longer appears to change within the desired level of resolution. Successive terms that become smaller and smaller have contributed less and less to the sum. Nevertheless, we cautioned in a footnote that, even though these terms decrease very rapidly in magnitude, there are so many of them that together they might still make a substantial contribution to the sum. However, mathematicians have shown that, because even the ratio of each term to the preceding one decreases sufficiently rapidly *in this instance*, there is no cause for concern. We therefore write the expression for the number e as follows:

$$e = \sum_{l=0}^{\infty} \frac{1}{l!}.$$

Here the symbol ∞, infinity, may now be interpreted as follows: Add as many terms as you wish, sufficient in number that any additional ones will, within your pre-set level of resolution, make no noticeable contribution.

Infinity is defined as "larger than any pre-set value," and we shall frequently link this pre-set value to a level of resolution. The fact that the successive terms decrease in magnitude and become negligible within our level of resolution as long as l is chosen large enough, is expressed as follows:

$$\lim_{l \to \infty} \frac{1}{l!} = 0 .$$

In Chapter XIV we found the following equality for the capital resulting at time t when an amount r_0 is compounded at an annual interest rate of a percent per year in intervals of δt:

$$r(t) = r_0(1 + 0.01\, a\, \delta t)^{t/\delta t} .$$

We then found the number e by letting the compounding interval δt become so small that within our level of resolution the graph of the function $r(t)$ would appear to be smooth. That process may be expressed mathematically as follows:

$$\lim_{\delta t \to 0} r(t) = r_0 e^{0.01 a t} .$$

Recall that the exponential function was characterized by a constant *relative* rate of change:

$$\frac{1}{r}\frac{\delta r}{\delta t} = 0.01\, a$$

Here the ratio $\delta r / \delta t$ represents the time-rate of change of the capital r in the time interval δt; even though that time interval itself may be very small, and the increase in capital during that small time period will be small as well, their ratio is proportional in this instance to the product of the capital and the rate of interest:

$$\frac{\delta r}{\delta t} = 0.01\, a r .$$

Thus, while the individual increments in capital and time may within our level of resolution become too small to measure, their ratio will still be meaningful as a measurable rate of change. We call this ratio the *derivative* of r with respect to t:

$$\frac{dr}{dt} \equiv \lim_{\delta t \to 0} \frac{\delta r}{\delta t} .$$

(Even though the notation $\frac{dr}{dt}$ appears to imply a ratio, the individual expressions dr and dt have no separate meaning.)

It is tempting, and common practice, to regard the derivative as an "instantaneous" rate of change. However, young Buckminster would object that such an instantaneous rate is inconceivable, because in a zero time interval there can be no change. Nevertheless, we observed in the example of compounding interest that decreasing the time interval between successive interest computations ultimately no longer had any appreciable effect, so that we could extrapolate the observations made in the tiniest measurable time interval into even smaller but no longer directly observable time intervals. Usually one can assume that in an interval so small that within our level of resolution it can no longer be measured, no surprise will occur. This assumption is by no means general, however; it is exactly what Buckminster Fuller objects to when he states that ultimately all structure is discrete. Extrapolation into immeasurably small intervals assumes an ultimate smoothness; it is an artifice, albeit a useful one, which ignores the hierarchy of structures at different levels.

Figure 15-1 (a) illustrates a smooth function, but Figure 15-1 (b) represents a function which at time $t = t_c$ abruptly changes its value, so that at the interval $t_c - \frac{1}{2}\delta t < t < t_c + \frac{1}{2}\delta t$ one can no longer say that decreasing the interval indefinitely will have no appreciable effect on the measured rate of change. We say that the function illustrated in Figure 15-1 (a) is continuous, but that the one in Figure 15-1 (b) has a discontinuity at $t = t_c$. A derivative cannot be defined at a point of discontinuity.

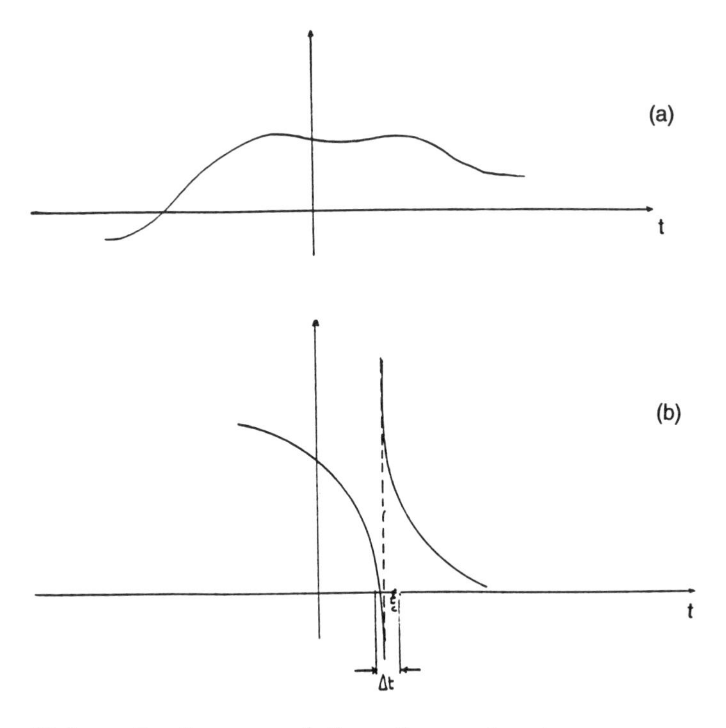

Figure 15-1 ***Continuous and discontinuous functions***

We have considered the rate of growth of capital with time and the rate at which the radial coordinate of the spiralling bugs decreases with respect to the angular coordinate. In the first instance time is called the independent variable, capital the dependent one. In the second example the independent variable is the angular coordinate, the dependent one the radial coordinate. We found that both are illustrations of the exponential function. Generalizing, we shall consider other functions as well, denoting the independent variable by x, the dependent one by y. The derivative of y with respect to x is defined in analogy with the definition given previously for the exponential function:

$$\frac{\delta y}{\delta x} = \frac{y(x+\delta x) - y(x)}{\delta x}; \; \frac{dy}{dx} \equiv \lim_{\delta x \to 0} \frac{\delta y}{\delta x}.$$

Here $y(x)$ stands for the value of the dependent variable when the independent variable has the value x; similarly, $y(x+\delta x)$ is the value of the dependent variable when the independent variable has the value $x+\delta x$. Let us test this definition against the already familiar exponential function, $y = e^{kx}$:

$$\begin{aligned} y(x+\delta x) &= e^{k(x+\delta x)} = e^{kx} \cdot e^{k\delta k} \\ y(x) &= e^{kx} \\ \frac{\delta y}{\delta x} &= \frac{e^{kx}(e^{k\delta x} - 1)}{\delta x} \end{aligned}$$

We can rewrite the definition of the number e in terms of the notation introduced in this chapter:

$$e \equiv \lim_{k\delta x \to 0} (1 + k\delta x)^{1/k\delta x}.$$

Hence $\lim\limits_{k\delta x \to 0} e^{k\delta x} = \lim\limits_{k\delta x \to 0} (1 + k\delta x)$:

$$\begin{aligned} \lim_{k\delta x \to 0} e^{k\delta x} - 1 &= \lim_{k\delta x \to 0} k\delta x \\ \frac{dy}{dx} &= e^{kx} \cdot k = ky. \end{aligned}$$

The conclusion that the derivative of the exponential function is proportional to that function itself is in accord with our finding in the previous chapter, and gives us confidence in using the same procedure to find the derivative of other functions.

One function which we shall need later on is $y = x^n$. As before:

$$\begin{aligned} y(x+\delta x) &= (x+\delta x)^n \\ y(x) &= x^n \\ \frac{dy}{dx} &= \lim_{\delta x \to 0} \frac{(x+\delta x)^n - x^n}{\delta x} \end{aligned}$$

Remembering the generalization of Pascal's triangle (Chapter III), $(x + \delta x)^n = x^n + nx^{n-1}\delta x + O(\delta x)$, where $O(\delta x)$ stands for the entire collection of terms containing powers of δx higher than 2. Therefore $\frac{dy}{dx} = nx^{n-1} + \lim_{\delta x \to 0} \frac{O(\delta x)}{\delta x}$. Since $O(\delta x)$ is a collection of terms containing powers of δx not lower than two, this collection will become negligible even when divided by δx, so that

$$\frac{d}{dx}x^n = nx^{n-1}.$$

Two additional properties of derivatives are useful to know about. In the first place, a constant by definition has the same value for all values of the independent variable, with the result that its derivative is necessarily zero. Secondly, if a is constant,

$$\begin{aligned}\frac{d(ay)}{dx} &= \lim_{\delta x \to 0} \frac{ay(x + \delta x) - ay(x)}{\delta x} \\ &= a[\lim_{x \to 0} \frac{y(x + \delta x) - y(x)]}{\delta x} = a\frac{dy}{dx}.\end{aligned}$$

In other words, the derivative of the product of a constant and a variable equals the product of that constant with the derivative of the variable.

XVI

Integrals and Logarithms

The radial coordinate of the gyrating bugs and capital compounded "continuously" are both examples of exponential functions, whose relative growth is constant:

$$\frac{dy}{dx} = ky\,. \qquad (16-1)$$

This expression is called a *differential equation*. Its solution is a function (in this case the exponential function) rather than a numerical value as is the case for an algebraic equation. The differential equation tells us something about the rate of growth of the function, a fact which may be even more useful than knowing the exact form of the function itself. As a matter of fact, certain analog computers called *differential analyzers* generate a function by using an electronic device called an *integrator*, which receives the derivative as input and produces the function as output. Figure 16-1 schematically represents the solution of equation 16-1: The output of the integrator is "fed back" after being multiplied by the constant k. The resulting feedback loop represents the differential equation 16-1: The derivative of y fed into the integrator is just the product of the constant k and the function y.

The device shown schematically in Figure 16-1 is called an integrator because the operation of finding a function when its derivative is known is

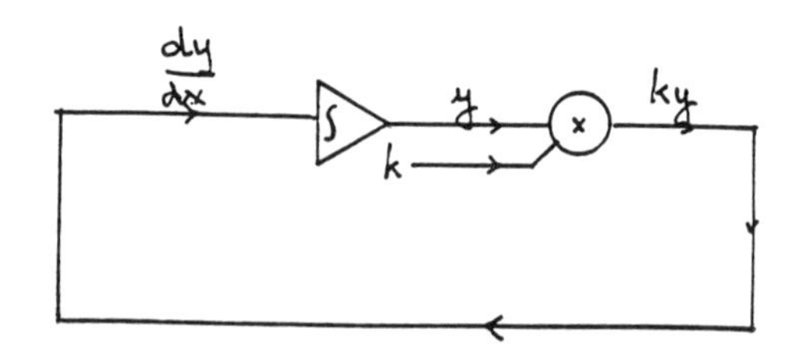

Figure 16-1 ***Analog computer element***

called *integration*. Integration is the inverse of differentiation. The *integral* of x^n is $x^{n+1}/(n+1)$, because, according to the rule derived in the previous chapter, the derivative of the latter yields the former.

The constant k in equation (16-1) is proportional to the interest rate in the case of compounded capital, and in the graphic representation as a spiral this constant determines the angle which the curve makes with the radius. However, knowing the value of k is not quite sufficient to compute the value of your capital or the position of the bugs at any given time. Indeed, you will need to specify your initial capital and the distance between the bugs at the beginning of their quest in order to know their future course. In point of fact, when the independent variable is time, then to solve, in other words, to integrate, a differential equation, you must give as many *initial conditions* as there are derivatives in the equation. Analogously, when the variables are coordinates, (in the examples of the bugs, the angle and radial distance), then every derivative in the differential equation necessitates a *boundary condition*, that is to say, a pair of values determining a point through which the curve is known to pass or at which the curve starts.

Recalling that if c is a constant, $\frac{d(cy)}{dx} = c\frac{dy}{dx}$, we realize that, whereas $y = e^{kx}$ is one solution of equation (16-1), multiplication of the exponential function by any constant will again produce a solution of that equation: Any function $y = ce^{kx}$ will be a solution. The constant c can only be determined once a boundary or initial condition is given. That condition may be expressed as follows: when $x = x_0, y = y_0$. Then:

$$y_0 = ce^{kx_0}, \quad \text{hence } c = y_0e^{-kx_0}, \quad \text{and } y = y_0e^{k(x-x_0)}.$$

Q: *RECALL THAT THE BUGS STARTED AT THE CORNERS OF A SQUARE HAVING UNIT EDGELENGTH. WRITE THE EQUATION FOR THE PATH FOLLOWED BY THE BUGS.*

We recall that in this instance $k = -1$, and that we called the independent variable measured in radians θ, the dependent one r. The initial condition is $r_0 = \frac{1}{2}\sqrt{2}$, when $\theta = 0$, so that the path of the bugs may be described by the equation $r = \frac{1}{2}\sqrt{2}e^{-\theta}$.

It might be useful to know how long one would have to compound capital to double its value, or through how large an angle the bugs would have to rotate to decrease their mutual distances by a given factor. This means that the role of the dependent and independent variables will have reversed. To express x in terms of y, we need a new function which is the inverse of the exponential function in the same sense that squaring is the inverse of taking a square root. Actually this "new" function is the logarithm; to appreciate this, we need to recall some properties of logarithms:

$$\begin{aligned} \log uv &= \log u + \log v \\ \log u^w &= w \log u \\ \log(1/u) &= -\log u\,. \end{aligned}$$

Apply these properties to the expression $y = ce^{kx}$:

$$\begin{aligned} \log y &= \log c + \log e^{kx} = \log c + kx \log e; \\ kx \log e &= \log y/c;\ x = \frac{\log y/c}{k \log e}\,. \end{aligned}$$

Usually, logarithms are used to the base 10. However, it is more convenient for our purposes to use the number e as the base, so that we can write $\log e = 1$. Logarithms to the base e are called "natural logarithms," and are denoted by the abbreviation ln. Hence, if

$$y = y_0 e^{k(x-x_0)}, \quad \text{then} \quad x = x_0 + \frac{1}{k} \ln y/y_0\,. \qquad (16-2)$$

We have thus related the exponential function to the more familiar logarithmic one; that the latter is more familiar than the former appears to be born out by the fact that the spiral curves generated by the bugs are commonly known as *logarithmic* spirals. Such spirals are representations of exponential functions (or logarithmic ones) in *polar coordinates*. Because they represent processes of constant relative growth rate, such spirals are quite common in nature. Figure 16-2 shows a cross section of the shell of the chambered nautilus. The chambers serve as flotation devices; the larger

Figure 16-2 ***Chambered nautilus***

its inhabitant, the larger the flotation device needed. Every year a larger chamber is added, the rate of increase in size being proportional to the size of the inhabitant, hence also of the previous chamber which kept it afloat. The appearance of logarithmic spirals in the nautilus shell is therefore not the result of some pre-programmed logarithm table in the nautilus, but a natural consequence of the growth mechanism and the structural function of the chambers.

It is easy to understand the expression "spiralling inflation," for the logarithmic spiral is a visual metaphor for exponential growth. Nevertheless, polar coordinates are not always the best frame in which to display the growth of capital; whereas times goes on indefinitely, the angle θ appears to return to its original value after every revolution (360°), so that for every angle there are many radial distances, each one revolution removed from its nearest neighbors. If one wants to show cyclical functions of times such as seasonal variations, then polar coordinates will be suitable, as one revolution could represent a whole year. To plot capital growth, however, it is more convenient to use *cartesian coordinates*, in which the independent variable (time in this instance) is plotted in the x-direction, and the dependent variable in the y-direction perpendicular to the x-direction.

Q: *PLOT THE PROGRESS OF THE BUGS IN CARTESIAN COORDINATES.*

Figure 16-3 shows two examples of an exponential function, one having $k > 0$, the other having $k < 0$. A third example of the expo-

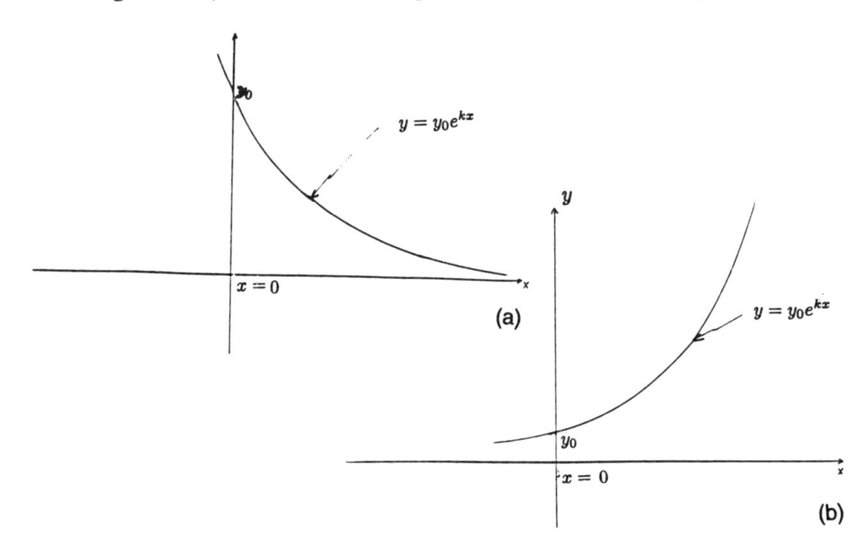

Figure 16-3 ***Exponentially (a) decreasing and (b) increasing functions in cartesian coordinates***

nential function in nature is radioactive disintegration. Here the rate at which the amount of an isotope decreases is proportional to the amount present. The curve corresponding to $k < 0$ in Figure 16-3 shows that this rate gradually becomes smaller, but never quite vanishes. We say that this curve approaches the x-axis *asymptotically*: It comes ever nearer to the axis but never intersects it.

Q: *FOR THE EXPONENTIALLY DECREASING PROCESS SHOWN IN FIGURE 16-3, DETERMINE HOW LONG IT WOULD TAKE FOR THE DEPENDENT VARIABLE TO DECREASE TO HALF ITS ORIGINAL VALUE. DO SO FOR A NUMBER OF DIFFERENT STARTING POINTS, AND COMPARE THE RESULTS.*

Let us set as our initial condition that at time $t = 0$ the amount of isotope present is y_0. Then the amount at any time is:

$$y = y_0 e^{kt} .$$

If we call the time at which only half of the initial amount is present $t = t_{\frac{1}{2}}$, then $\frac{1}{2} = e^{kt_{\frac{1}{2}}}$, and (remember that k is negative)

$$kt_{\frac{1}{2}} = -\ln 2 . \qquad (16-3)$$

When we calculate analagously how much time would pass before half of the original amount had further decreased to a quarter of the original, we find the same expression for $t_{\frac{1}{2}}$. It is called the *half-life* of the isotope; it is the time period over which half of the isotope disappears. It is characteristic of exponential decrease that the half-life is independent of the amount of isotope initially present. The larger the half-life, the slower the disintegration process is. Equation (16-3) shows that the half-life is reciprocally related to the constant k. Recall that the constant generally stands for the relative rate of change:

$$\frac{1}{y}\frac{dy}{dt} = k ,$$

and in the case of the compounded interest was proportional to the interest rate.

Equation (16-2) tells us that, if we were to plot $\ln y/y_0$ against x, we should obtain a straight line whose slope is k. Figure 16-4 (a) shows a sheet of semi-log graph paper; in Figure 16-4 (b) the curves of Figure 16-3 are replotted on tracing paper overlaid on such paper.[1] (On log-log paper *both* axes would be marked logarithmically.) Since exponential processes are so common, scientists frequently plot their data on semi-log paper; if the data lie on a straight line, then the process is indeed exponential, so

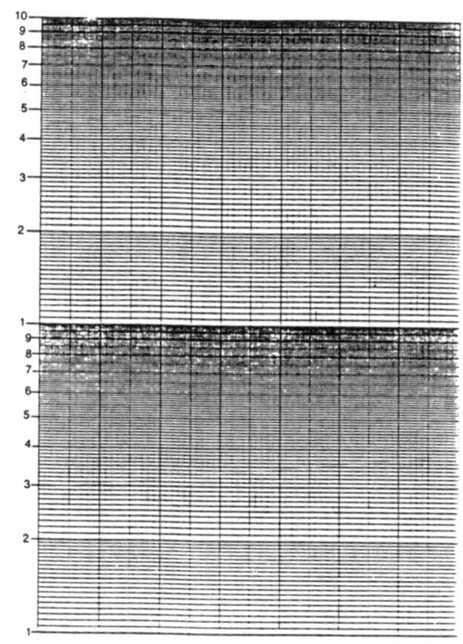

Figure 16-4 (a) ***Semi-log paper***

that the constant k and the half-life can be computed from the slope of the graph. A straight line is determined by two points; the coordinates of all other points are easily read off a straight-line graph or calculated by linear interpolation. For these reasons semi-log graph paper is very useful in connection with exponential processes.

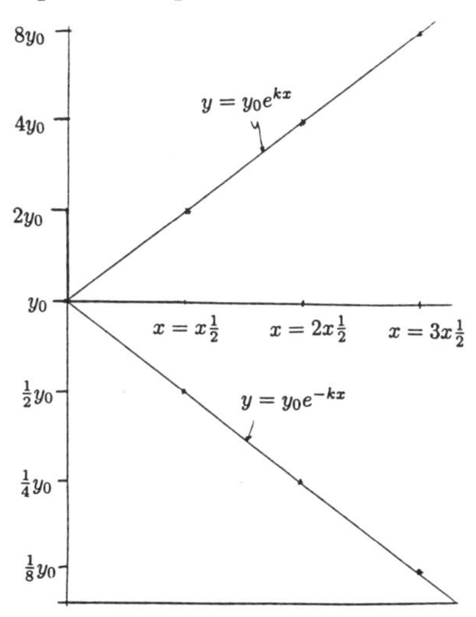

Figure 16-4 (b) ***Exponential functions plotted on semi-log paper***

Q: *FROM THE GRAPH ON SEMI-LOG PAPER, COULD YOU READ OFF THE VALUE OF THE INDEPENDENT VARIABLE FOR WHICH AN EXPONENTIALLY DECREASING VARIABLE REACHES THE VALUE ZERO?*

Note that on semi-log paper the values of the dependent variable decrease as the horizontal axis (the axis of abscissae) is approached. As the straight line representing the exponentially decreasing variable approaches the axis of abscissae, it would appear easy to determine when this line intersects the axis of abscissae, hence reach the point where the dependent value becomes zero. However, closer inspection reveals that the axis of abscissae may represent a small value of the dependent value, but not *zero*. In point of fact, the straight line representing the exponentially decreasing process will cross the axis of abscissae and continue on down. Every time the independent variable (time) increases by a given *amount*, the dependent variable will decrease by a given *factor*, and the value zero will never be reached because it is located infinitely far below the axis of abscissae on the semi-log paper.

Q: *PLOT THE TRAJECTORY OF THE BUGS ON SEMI-LOG PAPER. WHERE IS THE POINT WHERE THEY WILL MEET?*

NOTES

[1] When using semi-log paper, one should plot the dependent variable as a *ratio*. Usually it is convenient to assign $y = y_0$ to the value marked unity on the ordinate axis, so that y/y_0 is plotted as the dependent variable.

XVII

If you drop a ball from a given height, its velocity will increase at a constant rate called the acceleration of gravity:

$$\frac{dv}{dt} = g\,. \tag{17 – 1}$$

Integrating, we find that $v = gt$, assuming that at time $t = 0$, the moment at which the ball was dropped, the ball had zero velocity. Actually the velocity is itself the rate at which the altitude of the ball decreases:

$$v = -\frac{dh}{dt}$$

Since $v = gt$, $\frac{dh}{dt} = -gt$, and $h = h_0 - \frac{1}{2}gt^2$, where h_0 is the initial height from which the ball was dropped at time $t = 0$.

Equation (17-1) is correct only if air friction is neglected, which is strictly justified only *in vacuo*. As the ball speeds up, however, it will experience an increasing frictional force which opposes the motion, i.e., the acceleration process. To a first and very good approximation, this force is proportional to the velocity and acts in a direction opposite to that of the velocity. Therefore equation (17-1) has to be corrected for friction as follows:

$$\frac{dy}{dt} = g - ky\,. \tag{17 – 2}$$

This equation resembles that for the exponential function, so let us try a solution of the form $v = a + be^{-kt}$, where a and b are constants still to be determined. When we substitute this expression in equation (17-2), we find for the left hand side:

$$\frac{dv}{dt} = -bke^{-kt}\,,$$

and for the right hand side:

$$g - kv = g - ak - kbe^{-kt}\,.$$

The two are equal as long as $g = ak$, hence $a = g/k$, so that

$$v = g/k + be^{-kt}.$$

If the ball is dropped without initial velocity:

$$0 = g/k + b, \quad \text{so that} \quad b = -g/k, \quad \text{and :}$$
$$v = (g/k)\,(1 - e^{-kt}) \tag{17-3}$$

Q: *WHEN DOES THE VELOCITY REACH THE VALUE q/k?*

This function is illustrated in Figure 17-1. It is seen that ultimately velocity will approach a value $v = g/k$ asymptotically. When the frictional force is very small, then k is small, and the asymptotical velocity g/k will be very large. Figure 17-1 (a) illustrates a large frictional force, 17-1 (b) a small one.[1] Remembering the definition of e we recall that

$$e = \lim_{u \to 0} (1 + u)^{1/u}.$$

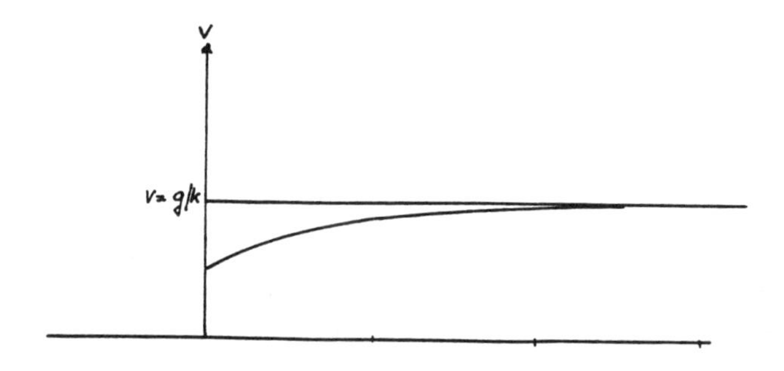

Figure 17-1 (a) ***Velocity of a body falling under the influence of gravity and friction: (a) High friction***

This means that for sufficiently small values of u, e^u approaches $1 + u$ arbitrarily closely. If we set $u = -kt$, then for sufficiently small values of kt, equation (17-3) approaches $v = gt$, which is the solution for the frictionless case described by equation (17-1).

The expression kt is small when both k and t are small. However, no matter how small the frictional force is, after a sufficiently long time the approximation $v = gt$ will no longer be valid: The velocity will then have attained such large values that the frictional force will become appreciable.

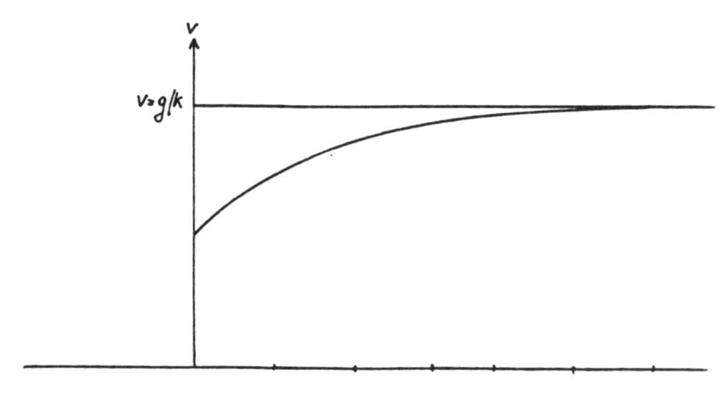

Figure 17-1 (b) ***Low friction***

Q: *DO YOU THINK THAT A FALLING BODY FALLS WITH CONSTANT VELOCITY OR CONSTANT ACCELERATION? WOULD YOU CONSIDER ONE OR THE OTHER A SUPERSTITION?*

Figure 17-1 has historical significance when we recall that the ancients believed that for free fall the velocity, not the acceleration, is constant. We see now that in air or another medium the velocity will initially increase at a constant rate, but will eventually become constant itself as a result of friction. For a sphere falling in a liquid this behavior is easily verified. It is observed, therefore, that the ancient belief is not entirely unjustified, but that in air the frictional effect will not be noticeable until quite high velocities are reached. The curves in Figure 17-1 follow, immediately upon release of the ball, the straight line corresponding to constant gravitational acceleration. The initial slope of all curves is the same, namely the acceleration due to gravity, g. Eventually these curves level off toward steady velocity; this velocity is proportional to the acceleration due to gravity and inversely proportional to the friction constant k. Experiments involving dropping objects from towers span such small time intervals that the effect of friction is negligible, but we know that objects dropped from stratospheric altitude will ultimately experience so much friction that they burn up in our atmosphere.

Q: *IS THE INITIAL SLOPE OF FIGURES 17-1 (a) AND 17-1 (b) THE SAME? (REMEMBER THE DIFFERENT TIME SCALES!)*

We have considered exponential functions generated by the bugs, by compounding interest, and by radioactive disintegration, as well as the fall of an object subject to gravity and friction. In each of these instances the rate of change of the dependent variable is determined by the dependent variable itself and by certain constants such as the interest rate, frictional force or half-life. The independent variable did not explicitly play a role in determining this rate of change: Interest is calculated on the basis of the amount of money in the bank, not on the date on the calendar or the season of the year. Of course, the interest does increase with time, but only because the capital itself changes. It is conceivable that the interest *rate* itself will change with time; in that case the growth rate will be affected. However, even in such a case, it is more likely the overall supply of cash that will determine the interest rate rather than the date on the calendar.

There are processes in which rate of change is determined explicitly by the independent variable. If you have royalties deposited monthly into your account, then your balance will be determined by the calendar: Every month on a given day it increases by a variable amount, which is not determined by how much is already in your account, but by the number of books sold in that period. The growth rings on a tree will show variations depending on variations in the climate from year to year. Generally, then, growth rate will depend on both the dependent and the independent variables as well as on certain constants; the differential equations considered so far are special cases of a more general form

$$\frac{dy}{dx} = f(y, x, k_1, k_2, \ldots),$$

where k_1, k_2, etc., are independent of x and y.

Those functions whose rate of change does not explicitly depend on the independent variable are called *growth functions*.[2] The exponential function and the related function determining the velocity of an object subject to gravity and friction are examples of growth functions. We shall now consider some additional growth functions.

In the 1950s following the success of Sputnik, there was an apparent exponential increase in the number of scientists in the USA; as a result it was said that by the end of the century there would be more scientists than people. One might similarly speculate what would happen if everyone having an account at a given bank were to leave her or his account to accumulate compounded interest: Some money would have to come from outside to sustain the interest payments. Both examples are obviously absurd, for we know perfectly well that the growth of a field like science will be limited by the supply of people able to enter it, and that interest rates will be adjusted according to the supply of lenders or borrowers. Accordingly, an adjustment must be made in the "constant" k in order to provide a model which may be extrapolated a long time ahead. If y denotes the total amount of capital deposited in a bank, then a critical value y_c must be set by the bank so that when y begins to approach that critical value, the interest rate will decrease, and the growth rate of invested capital will level off toward zero. Similarly, as the number of scientists increases, the fraction of the population entering science will level off as society will become unable to support more than a certain fraction of itself as scientists.

Certain chemical reactions are speeded up by the presence of substances called *catalysts*. Occasionally, one of the products of a reaction will itself catalyse a reaction; reactions for which this happens are called *autocatalytical*. As a result the reaction will speed up as more of the catalytical substance is produced. However, the speed of the reaction is also determined by the amount of the original substance present, and as that becomes used up the reaction will slow down.

Growth functions are solutions of differential equations of the form

$$\frac{dy}{dx} = f(y) \qquad (17-4)$$

where $f(y)$ is independent of x.

The examples considered so far, the amorous bugs, compounded interest, the body falling in a frictionous medium, have all been analyzed using a function $f(y)$ of the first degree ("linear") in y; in a frictionless medium $f(y)$ was even of degree zero, i. e., it was independent of y as well as of x. The right hand side of equation (17-2) vanished when $v = g/k$; accordingly, dv/dt will also vanish and v will no longer change if it attains that value. Equation (17-3), a solution of equation (17-2), shows that indeed v approaches the asymptotic value g/k, but never reaches it because its rate of change decreases as v comes closer to its asymptote g/k.

In the more general case represented by equation (17-4), there would be an asymptote for every value of y which causes $f(y)$ to vanish. Let us consider a quadratic form of $f(y)$, say one which is positive in the interval $a < y < b$ and vanishes for $y = a$ and for $y = b$:[3]

$$f(y) = k(y-a)(b-y).$$

Then

$$\frac{dy}{dt} = k(y-a)(b-y), \qquad (17-5)$$

Before we solve differential equation (17-5), in other words, express y as a function of t, it is important to realize that the differential equation itself may give us insights into the appearance of the solution. We have already noted that the growth rate of y vanishes for $y = a$ and for $y = b$. Since the independent variable t plays no *explicit* role in the growth functions, we can gain some interesting insights into the behavior of growth functions before knowing explicitly how y depends on t, by plotting the growth rate, dy/dt, against y itself. Figure 17-2 (a) exemplifies the case of constant acceleration: dy/dt is constant. In Figure 17-2 (b) we have the case of

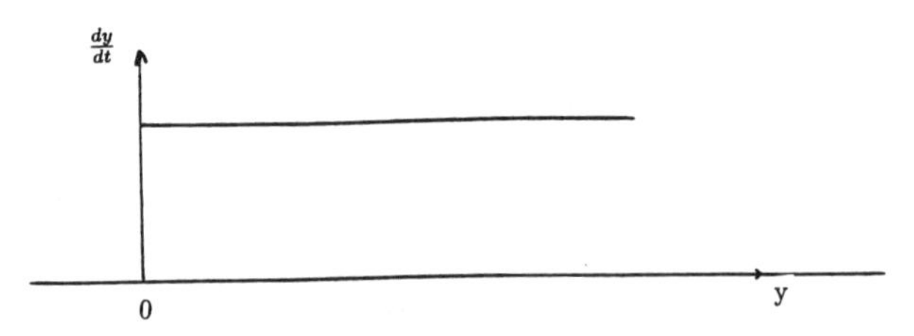

Figure 17-2 (a) ***Rate of change of y plotted against y: (a) Constant acceleration***

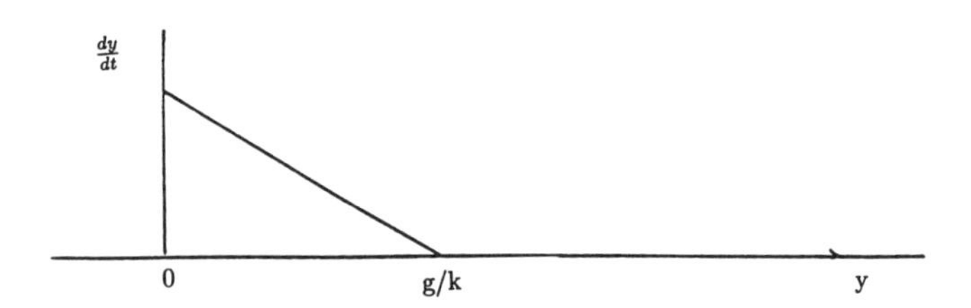

Figure 17-2 (b) ***Gravity and friction***

fall subject to friction: The change of velocity is a linear function of the velocity itself, $g - kv$. At the initial moment $t = 0, v = 0$, and the velocity increases at the rate determined by gravity. The graph shows, however, that the rate of increase of the velocity (the acceleration) itself decreases, and would stop if a value $v = g/k$ were reached, which determines the asymptote.

We may similarly plot dy/dt as given by equation (17-5); the result is a parabola whose axis is oriented vertically. We decided to consider cases when dy/dt is positive in the interval $a < y < b$; the parabola then will be situated as shown in Figure 17-2 (c). (We shall learn presently that it is easy to generalize to other instances once the principles are understood for the present example.)

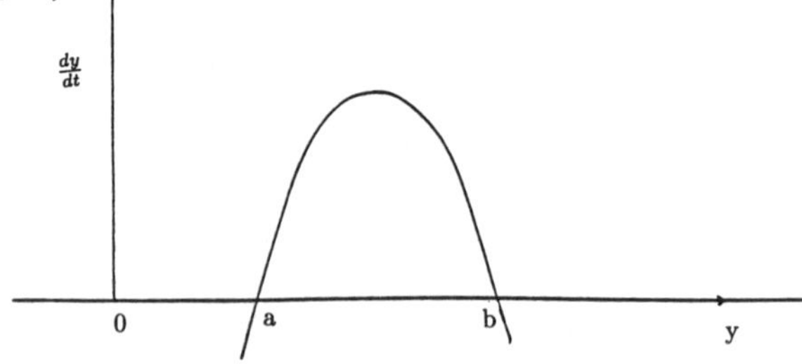

Figure 17-2 (c) ***Quadratic dependence of rate change on the dependent variable***

Q: *FOR WHAT VALUE OF* y *IS THE GROWTH RATE MAXIMUM?*

We note that growth ceases at the asymptotic values $y = a$ and $y = b$; we also see that as y increases in the interval $a < b, dy/dt$ itself reaches a maximum value halfway between $y = a$ and $y = b$, at $y = \frac{1}{2}(a + b)$, then decreases to zero again. Without even solving equation (17-5), we may sketch the course of y as a function of t from just these data (Figure 17-3).

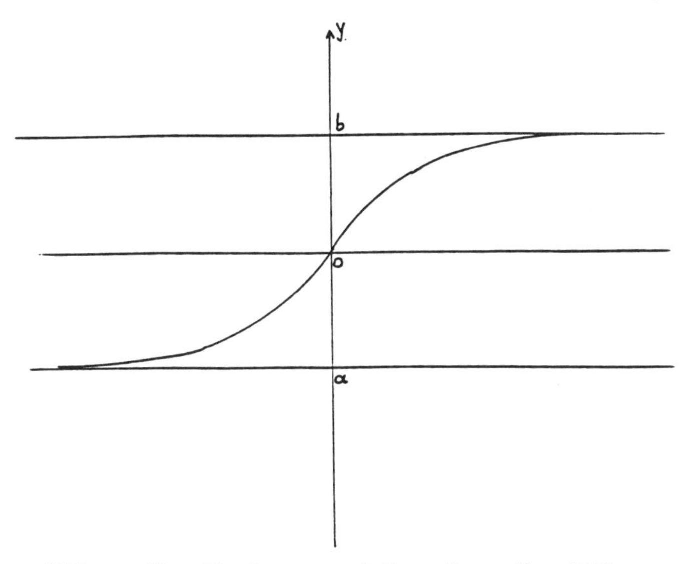

Figure 17-3 ***Graphical representation of equation 17-5***

Equation (17-5) may be rewritten:

$$\frac{1}{(y-a)} \cdot \frac{1}{(b-y)} \frac{dy}{dt} = k\,.$$

Let us write

$$\frac{1}{(y-a)} \cdot \frac{1}{(b-y)} = \frac{P}{(y-a)} + \frac{Q}{(b-y)},$$

with the values of P and Q to be determined. Multiplication of both sides of this last equation by $(y - a)(b - y)$ gives

$$1 = P(b-y) + Q(y-a), \quad \text{hence}$$
$$1 = (Q-P)y + (Pb - Qa)\,.$$

Of this last equation the left hand side is independent of y, therefore the right hand side must be as well. This is possible only if the coefficient of y on the right hand side vanishes, hence $P = Q$. The last equation then becomes:

$$1 = Q(b-a), \quad \text{hence} \quad Q = 1/(b-a), \quad \text{and} \quad P = Q = 1/(b-a)\,.$$

$$\frac{1}{(y-a)} \cdot \frac{1}{(y-b)} = \frac{1}{(b-a)}\left[\frac{1}{(y-a)} + \frac{1}{(b-y)}\right], \quad \text{and}$$
$$\frac{1}{(y-a)}\frac{dy}{dt} + \frac{1}{(b-y)}\frac{dy}{dt} = k(b-a)(5)\,.$$

Recalling that the derivative of a constant is zero, we see that

$$\frac{dy}{dt} = \frac{d(y-a)}{dt} = \frac{-d(b-y)}{dt},$$

We have found that for any function $z(t) : \frac{1}{z}\frac{dz}{dt} = \frac{d}{dt}(\ln z)$. Therefore

$$\frac{d}{dt}[\ln(y-a) - \ln(b-y)] = k(b-a);$$
$$\text{since} \quad \ln(y-a) - \ln(b-y) = \ln[(y-a)/(b-y)],$$
$$\ln[(y-a)/(b-y)] = k(b-a)t + \text{constant},$$

where the constant will be determined when the initial value of y is determined, that is to say, when we decide which value y assumes at the moment which we choose to call $t = 0$. This moment can be chosen quite arbitrarily: Its choice is analagous to setting a stopwatch to zero at a convenient moment to run an experiment. It will be convenient to choose the moment $t = 0$ such that the constant vanishes:

$$\ln[(y-a)/(b-y)] = k(b-a)t\,. \qquad (17-6)$$

Q: *WHAT IS THE VALUE OF* y *WHEN t=0?*

At time $t = 0, \ln[(y-a)/(b-y)] = 0$. Since $\ln 1 = 0$, then at time $t = 0, y - a = b - y$, or $y = \frac{1}{2}(a+b)$. In other words, the moment $t = 0$ was chosen just when y has a value half-way between the asymptotic values $y = a$ and $y = b$, which, we noted above, is when the growth rate is itself maximal.

Equation (17-6) can be rewritten:

$$y = \frac{be^{kbt} + ae^{kat}}{e^{kbt} + e^{kat}}\,. \qquad (17-7)$$

To graph this expression, note that an exponential function becomes very large (exponential growth is very fast growth!) when its argument becomes large, equals unity when its argument equals zero, and becomes nearly zero when its argument becomes very negative. As we let the independent variable t become very large, the exponential function e^{kbt} will outdistance e^{kat}, because $b > a$. As a result, in both the numerator and denominator of equation (17-7), the second term will become negligible compared with the first; when they are discarded, the exponential functions will cancel out, so that y approaches the asymptotic value b when t becomes very large, just as we surmised by inspection of Figure 17-3. Conversely, when t becomes very negative, the first terms in both the numerator and denominator of equation (17-7) go to zero more rapidly than the second terms, and therefore may be discarded. As a result now the second terms are retained, the exponential functions once more cancel out, so that for very negative values of t, y approaches the asymptotic value a, as had also been surmised.

The function given by equations (17-6) and (17-7) and defined by differential equation (17-5) is called a *sigmoid* function. It is an important as the exponential function, and deserves to be as well known. It is a metaphor for processes which initially grow very slowly, then pick up speed, and ultimately saturate. Derek de Sola Price[4] called attention to this type of behavior after studying a wide range of phenomena such as the growth in the number of universities in Europe during the Middle Ages and Renaissance. Price, however, regarded sigmoidal processes as consisting of three phases, one of initial slow growth, the next one characterized by rapid growth, and a final saturation phase characterized by a decreasing growth rate. We have demonstrated here, however, that the process is a continuous one, that the three "phases" flow from one into another without any outside influence, because the sigmoid is a growth function whose rate of change is not explicitly determined by the independent variable, time. The sigmoid is symmetrical around $t = 0$: Rotation around the point $t = 0, y = \frac{1}{2}(a+b)$ through 180° brings the curve into coincidence with itself. For $t < 0$ the rate of growth increases with time; after reaching a maximum value at $t = 0$, the rate levels off.

The factor $(y - a)$ in the right hand side of equation (17-5) represents the growth stimulated by what has already been produced, in analogy to the compounding of interest, or the speeding up of an autocatalytical reaction when the product also catalyses, hence speeds up the reaction. The factor $(b - y)$, however, represents the slowing down due to saturation: Interest rates might decrease as a result of a surfeit of savings, or the autocatalytical reaction might slow down as the original reagent begins to run out.

Equation (17-7) may be rewritten as follows to facilitate examining the behavior of y for very negative values of t, by dividing both numerator and denominator by e^{kat}:

$$y = \frac{b \cdot e^{k(b-a)t} + a}{e^{k(b-a)t} + 1}.$$

Long division tells us

$$1/[e^{k(b-a)t}] = 1 - e^{k(b-a)t} + e^{2k(b-a)t} - \dots.$$

For very negative values of t, $e^{k(b-a)t} << 1$, so that only the first two terms resulting from the long division will be significant. Therefore, for $t << 0$, we may use the approximation

$$y = [be^{k(b-a)t} + a][1 - e^{k(b-a)t}].$$

Multiplying the two factors on the right hand side of this equation and again discarding higher powers of $e^{k(b-a)t}$, we find the approximation

$$y = a + (b-a)e^{k(b-a)t}.$$

Thus we note that the sigmoid curve will, in its very early stages, be indistinguishable from a curve rising exponentially from its initial asymptotic value. Accordingly, the leveling typical of sigmoidal behavior would be difficult to predict in the initial growth phase.

NOTES

[1] Note that the time scales in Figures 17-1 (a) and 17-1 (b) were chosen differently: The same time interval occupies half as much distance along the time axis of Figure 17-1 (a) as along that of Figure 17-1 (b).

[2] Loeb, A. L.: *Synergy, Sigmoids and the Seventh-Year Trifurcation*, The Environmentalist, **3**, 181–186 (1983), reprinted Chrestologia, **XIV**, #2, 4–8 (1989).

[3] This expression is equivalent to the more familiar polynomial of the second degree, for multiplying the factors together produces

$$f(y) = ky^2 - k(a+b)y + kab.$$

[4] de Sola Price, D. J.: *Measuring the Size of Science*, Proc. Israel Acad. Science and Humanities, **4**, 6 (1969).

XVIII

Sigmoids and the Seventh-year Trifurcation, a Metaphor

Because the sigmoid is such an important function, it would be useful to put it in a more universal form, independent of parameters such as a, b and k, which are specific to the particular system under consideration, such as interest rate, amounts of reagents, half-lives, etc. We can scale the variables in equation (17-5) as follows. Instead of y we shall define a variable z which equals $+1$ when $y = b$, and which equals -1 when $y = a$; its aymptotic values would be $z = -1$ and $z = +1$. We write $z = Ky + L$ and will determine K and L such that z has the desired asymptotes:

$$\begin{aligned} -1 &= Ka + L \quad \text{and} \quad +1 = Kb + L, \quad \text{hence} \quad K = 2/(b-a), \\ L &= (a+b)/(a-b), \quad \text{and} \\ z &= (2y - a - b)/(b-a), \; y = \frac{1}{2}[(b-a)z + a + b]. \end{aligned}$$

Substitute in equation (17-5):

$$\frac{dz}{dt} = \frac{1}{2}k(b-a)(1+z)(1-z).$$

Finally, we can define a new independent variable $x = \frac{1}{2}k(b-a)t$, putting the differential equation in a universal form:

$$\frac{dz}{dx} = (1+z)(1-z). \qquad (18-1)$$

The solution to this form of the equation can be written directly by setting $k = 1, y = z, x = t, a = -1, b = 1$ in equation (17-7):

$$z = (e^{x} - e^{-x})/(e^{x} + e^{-x}). \qquad (18-2)$$

So far we have considered only the range $a < y < b$, corresponding to the range $-1 < z < +1$. Values of z outside this range can be found easily as the result of the following surprising property of equation (18-1):

Theorem 18-1: If z is a solution of equation (18-1), then so is $1/z$.

This property is proven by direct substitution of $u = 1/z$ in equation (18-1), remembering that, if $u = 1/z$, then $dz/dx = -z^{-2}du/dx$, so that

$$\frac{du}{dt} = (1+u)(1-u)\,.$$

This equation is identical in form to equation (18-1), from which it was derived. As a result we find that for $z > 1$ and $z < -1$, equation (18-1) has the solution

$$z = (e^x + e^{-x})/(e^x - e^{-x})\,. \qquad (18-3)$$

Figure 18-1 shows the entire range of solutions of equation (18-1), the sigmoid in the range $-1 < z < +1$, and its reciprocal outside that range. It will be interesting to speculate what the meaning of the reciprocal sigmoid might be.

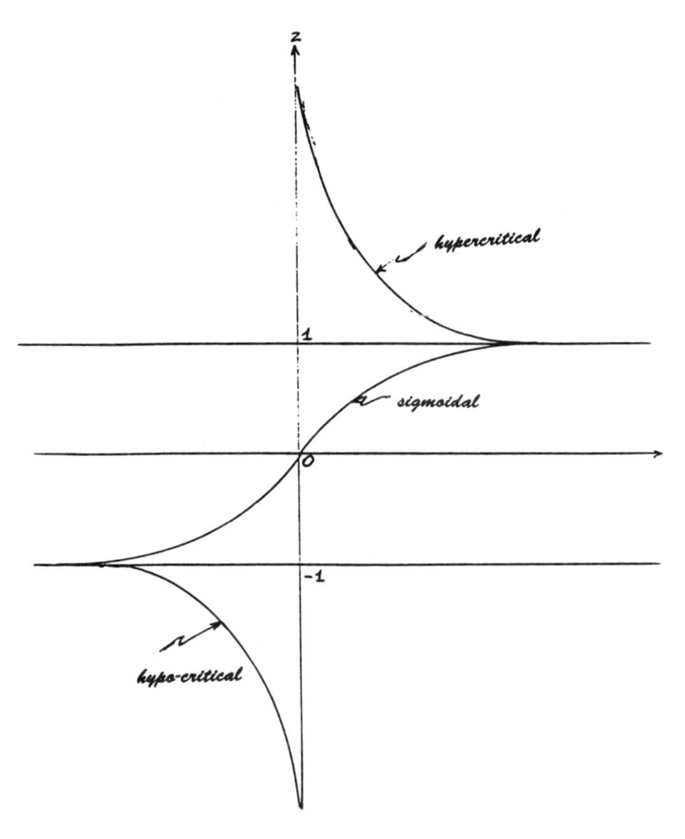

Figure 18-1 *Solution of equation (18-1)*

Figure 18-1 shows us that the values $z = -1$ and $z = +1$ are critical. If z is just a tiny bit larger than -1, then it will experience positive growth, but if it is just a bit smaller than -1, it will decrease, its rate of decrease growing catastrophically so that the value of z will go off the scale as x approaches zero. Whether z experiences sigmoidal growth or catastrophic decline is determined critically by its value when x is extremely negative. The behavior of this growth function may be considered metaphoric for what happens to many new enterprises: Some will experience growth, whereas others will lose money, accumulate debts, and fail precipitously. The initial difference between the two courses is very small but determines the future critically. Since the branch of the growth curve corresponding to $z < -1$ results from z having a less than critical value, we shall call it the "hypo-critical" branch, distinguishing it from the sigmoidal portion of the growth curve.

This critical division of the growth curve into separate branches qualitatively differs from the behavior of the growth curves corresponding to a function $f(y)$ of degree zero or one. Since exponential growth, corresponding to a first-order function $f(y)$, has entered the vernacular in our culture, it behooves one to understand as well the growth pattern corresponding to a quadratic $f(y)$.

For values of z near $+1$ the situation is the opposite of that near -1. It is true that, just as when z is near -1, the growth rate is negative when z is outside the range $-1 < z < +1$, and positive when inside that range, but whether z is less than or greater than $+1$, the growth rate goes to zero with increasing x, and two branches of the curve approach $z = +1$ asymptotically. Thus while $z = -1$ is an unstable value, from which the two branches diverge, $z = 1$ is stable, two branches converging asymptotically. Since $z = 1$ is also a critical value where the growth rate changes sign (cf. Figure 17-1), we shall call the branch corresponding to $z > 1$ the "hyper-critical" branch. It, too, may be used as a metaphor: Who has not encountered the over-qualified individual who, in a less than challenging situation, settles into a comfortable no-growth pattern?

The discovery that equation (18-1) has not only the sigmoidal solution but also the reciprocal has thus led us to the conclusion that there are three potential growth patterns corresponding to the quadratic $f(y)$ which vanishes for two distinct values of y, namely the catastrophic hypo-critical one, the sigmoidal one, and the hyper-critical one. The passage from the hypo-critical to the hyper-critical branch when the independent variable x passes through zero is a typical example of a discontinuous function: When x is very close to zero in equation (18-3), e^x may be approximated by $(1 + x)$, and e^{-x} by $(1 - x)$, hence their difference in the denominator by $2x$, with the result that z will be a very large positive number when x is positive, but a very large negative one when x is negative.

Although we have denied the existence of three distinct phases in the sigmoidal growth function, it is easy to see why three phases, initial and final ones characterized by very slow growth and a middle one of fast growth, would be perceived. To begin with, the rate of growth itself goes through a maximum: The sigmoidal curve experiences inflection at $x = 0$, when its slope ceases to increase and begins to level off to zero. (In this vicinity the sigmoid approximates a straight line.) In the second place, although an immense range of x is required for z to pass from very near -1 to very near $+1$, most of that change actually occurs within a relatively narrow range of x near $x = 0$. The half-life of the negative exponential function e^{-x} was defined as the increase in the independent variable necessary to halve the value of this exponential function, namely $\ln 2$, which is less than unity. Equation (18-1) shows us that over that rather narrow range of x, z changes from -0.6 to +0.6, or by 60% of its entire range; $\ln 2$ is therefore also a characteristic parameter of sigmoidal growth, with which we might associate the apparent range of rapid growth. When we recall the definition $x = \frac{1}{2}(b - a)t$, then we find that the time interval between the moment of maximum growth rate and the moment when the dependent variable reaches 60% of its asymptotic value is given by T:

$$T = 2 \ln 2 / k(b - a) .$$

This equation tells us that for large values of the parameter k and for a large difference between the values of t for which $f(y)$ vanishes, T is small, i.c., the duration of "rapid growth" is short, and the growth curve is steep in that range. The combination $k(b - a)$, hence the time constant T, is a characteristic parameter of sigmoidal growth, determining how wide the so-called middle range of rapid growth will be and how far the asymptotic values are apart.

Many readers should be familiar with the biblical story of the seven fat and the seven lean years, with the expression "the seven year itch," or with the notion that one should have a career change every seven years. The sigmoidal-growth model could be a metaphor for phases in a human life or career having a time constant of the order of seven years. In a culture conditioned by the expectation of exponentially growing grand national products, it is good to realize that the inclusion of a quadratic term in $f(y)$ will imply the likelihood of asymptotic leveling off. (The inclusion of even higher powers of y in $f(y)$ will simply add more asymptotes to y. The hyper- and hypo-critical phases would then also become sigmoidal, but having a negative slope, implying "redemption" at the end of a hypo-critical phase.) Whereas the exponential function might have been metaphoric for an America having a frontier, the sigmoid may represent a metaphor for a more mature society having no more room for indefinite expansion. We were familiar with "periods" in history or in an artist's career: Gothic styles, the Baroque, Art Deco, a painter's Blue Period. The question is: What happens when the top of the sigmoid is reached?

In Figure 18-2 we have a diagram of potential model. The author has elsewhere[1] presented some fictional scenarios to illustrate critical points in human or organizational careers for which this model could be a metaphor. Two growth functions have been superimposed such that the upper asymptote of the one coincides with the lower one of the other, with the points of inflection of the two sigmoids somewhat more than an interval T apart along the t-axis. Both growth curves have the same values of k and of $(b-a)$. The hypo- and hypercritical as well as the sigmoidal growth curves are displayed. In this model we observe that upon traversal of the lower left sigmoid a trifurcation is encountered: the sigmoidal and hyper-critical branches of the lower left growth curve approach a common asymptote, whereas the middle growth function offers a new sigmoid and a hypo-critical branch.

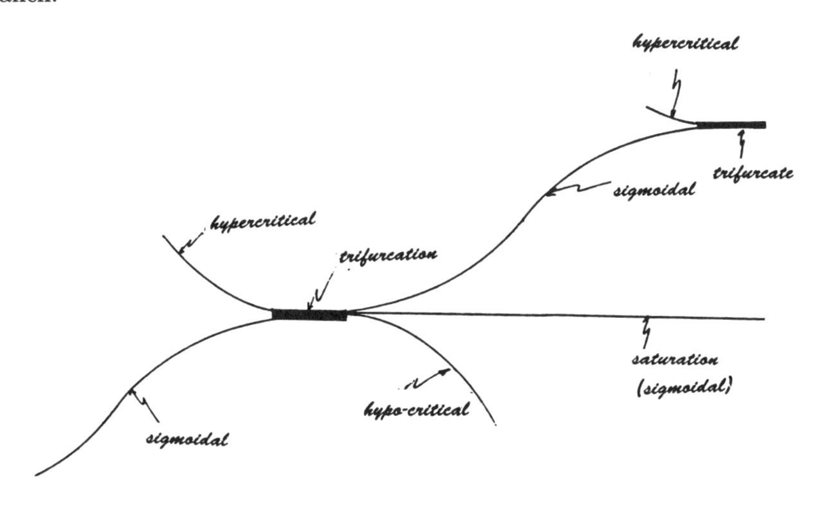

Figure 18-2 *Trifurcation model*

It would appear that after successful traversal of a sigmoid curve one could transfer to a new sigmoid function, with the inherent risk of landing on the hypo-critical rather than the sigmoidal branch, or one could remain on the saturation portion of the old function without further growth. The reader will not have great difficulty in recognizing examples of each of these three eventualities. We could recall Handel exploring the oratorio as a new form when his operas had saturated their popularity in London. By comparison we can think of the natural scientist who has scored an early triumph and then never equals that achievement, or of a failed career change. In point of fact, it is remarkable that artists are usually creative until a ripe old age, albeit changing their style periodically, whereas natural scientists are reputed to have scored early or not at all; apparently natural scientists have fewer opportunities than artists to transfer to a new growth function.

Equation (17-5) has, on its right side, two factors: $(y-a)$ and $(b-y)$. The growth rate is proportional to their product. In Chapter XVI, Figure 16-1, we showed how a differential equation is solved by analog computer, both factors being fed back into the integrator. Figure 18-3 gives the feedback model corresponding to equation (18-3).

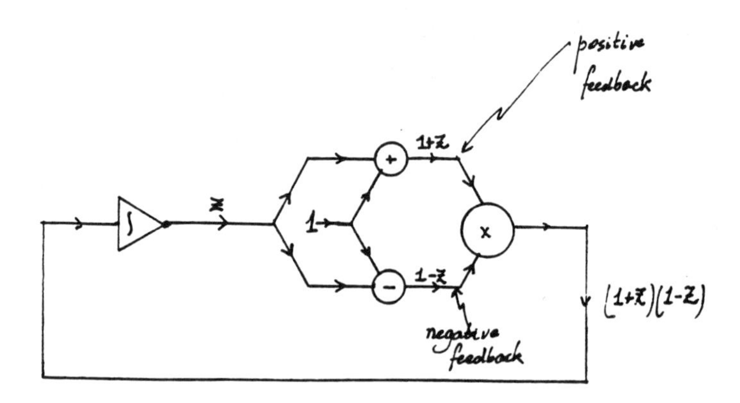

Figure 18-3 ***Feedback model for equation (18-1)***

In the range $a < y < b$ both factors produce positive feadback. In the initial sigmoidal phase the positive feedback due to $(y-a)$ is still small, and in the final phase the feedback due to $(b-y)$ keeps the growth rate small; when y is near $\frac{1}{2}(a+b)$ the two feedback mechanisms work together to maximum advantage.

The term "feedback" is used in connection with both cybernetics and behaviorism. A common example of cybernetic feedback is furnished by the thermostat: When the temperature exceeds a previously set standard, the furnace will stop until the temperature falls below the standard, at which moment the furnace will relight. The feedback is negative: Too high a temperature will result in cooling, too low a one in heating. The integrator used in analog computers to solve differential equations is a cybernetic device: The sigmoid function is traced out electronically such that its growth rate remains exactly proportional to the product of the two factors produced by the integrator and its associated adder, subtractor and multiplier.

B.F. Skinner showed that behavior may be modified by immediate attention to the result of that behavior. As in the case of cybernetic systems, a feedback loop is established relating the result of the behavior to the behavior itself, and in turn affecting that behavior. The analogy between cybernetic and behavioral feedback becomes important when we use the sigmoid as a metaphor for the behavior of organisms. When an entrepreneur starts a business, she may start with a very small investment, waiting to see whether she will attract customers. If sales go well, she will invest some of her profits in more varied stock, and attract more customers, producing more profits, etc. (Slow initial growth followed by acceleration of growth.) However, eventually she will discover that she will need to hire additional help, with the result that overhead will increase, but she will expand until she has exhausted all the potential customers, and no new ones will arrive. At that point an equilibrium will be reached, and the business will continue but no longer expand: The sigmoid will have reached its asymptote. Initial positive feedback produced increasingly rapid expansion, but her limited ability to manage a growing business and a limit to the number of potential customers eventually led to saturation. At this point she may choose to continue the business with the existing staff, or she may enter a new sigmoid, say by adding a mail-order component.

On the other hand, she might initially not have attracted enough customers to make a profit; rather than order additional stock, she would not even have been able to meet expenses. Unable to pay her rent she would have to borrow funds; interest payments would increase her overhead, and eventually her credit rating would suffer. Negative feedback would build up: The business is hypo-critical. Either eventuality is contingent on the feedback received from customers.

The question whether an artist can operate in complete isolation is controversial. The painter who keeps painting, accumulating work which nobody buys or even sees, is certainly rare. Most painters are affected by their success or lack thereof, and feedback from sales or reviews will influence artists. Some are so successful that they cannot change their style: They follow a sigmoid to saturation, producing in the same style over and over again. Others will experiment with something new, attempting new sigmoids, experiencing feedback which will guide their new ventures. Some artists start off in isolation, but will find support among like-minded colleagues, who will eventually group together, exhibit together, and attract more attention than each could have separately. Their productivity will grow; perhaps they will even define a style such as the impressionists did in the late nineteeth century. Eventually, however, viewers will find other interests, the artists will themselves find other styles or die, and the style will come to an end.

It is evident that feedback plays an important role in growth processes, and it is no coincidence that the simulation of growth processes on a differential analyzer uses cybernetic feedback techniques. The concept of feedback is so important in contemporary thinking that it may well supplant the search for cause and effect[2]: It does not matter at all which came first, the chicken or the egg. In cybernetics and behaviorism, effect is cause and cause is effect.[2,3]

Notes

[1] Loeb, A. L.: *Synergy, Sigmoids and the Seventh-Year Trifurcation*, The Environmentalist, **3**, 181–186, reprinted in: *Chrestologia*, **XIV**, #2, 4–8 (1989).

[2] Loeb, A. L.: *On Behaviorism, Causality and Cybernetics,* Leonardo, **24**, 299–302 (1991).

[3] Loeb, A. L.: *Can Renaissance Man Survive in a Competitive Culture?*, International J. for Social Education, 24–40 (1992).

XIX

Dynamic Symmetry and Fibonacci Numbers

Draw an agreeable looking rectangle: Choose whatever dimensions you like, but avoid drawing a square. Going around the rectangle clockwise, label the vertices A, B, C, and D respectively, taking care that the distance between A and B is greater than that between B and C. Then draw diagonal AC, and drop a line perpendicularly from B onto AC, extending this perpendicular until it intersects CD at a point which you will label E. Draw a line through E parallel to BC, and label its intersection with AB F.

Q: *YOU HAVE NOW DIVIDED YOUR ORIGINAL RECTANGLE INTO TWO OTHER RECTANGLES. CAN YOU FIND ANY SIGNIFICANT RELATIONSHIP BETWEEN ANY TWO OR ALL THREE OF THESE RECTANGLES?*

From the construction we note that, since BE is perpendicular to AC, and BC is perpendicular to AB, angle BAC equals angle CBE. Rectangles ABCD and BCEF are similar, so that **AB/ BC = BC/CE**, where bold type indicates the *length* of the line denoted (Figure 19-1).

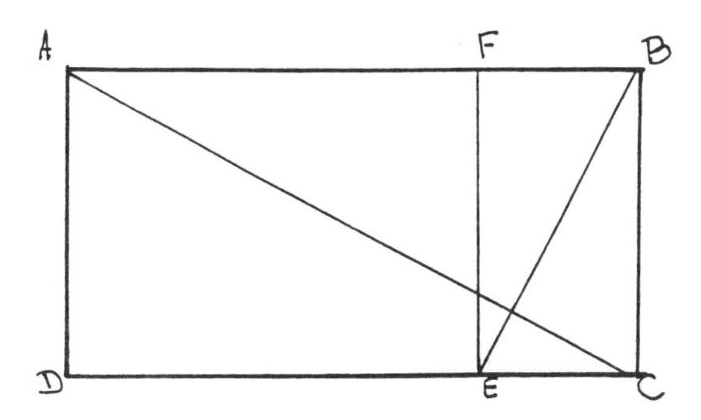

Figure 19-1 ***Special division of a rectangle***

The construction may now be repeated: For the rectangle BCEF, the diagonal BE is already in place, as is the perpendicular dropped onto it from C. All we need to do is to label its intersection with EF, which we call G, and then to draw GH parallel to EC (Figure 19-2). The new rectangle CEGH has its sides in the same proportion as did the original ABCD and BCEF. The construction may be continued indefinitely: Each time the most recently generated rectangle is subdivided into a rectangle geometrically similar to the original one and a second one which has been given the name of *gnomon*. Thus rectangle AFED is the gnomon of rectangle ABCD, BHGF that of BCEF, etc.

A process such as we just described is called *recursive*: Each successive rectangle is subdivided into one similar to itself and a gnomon by exactly the same procedure.

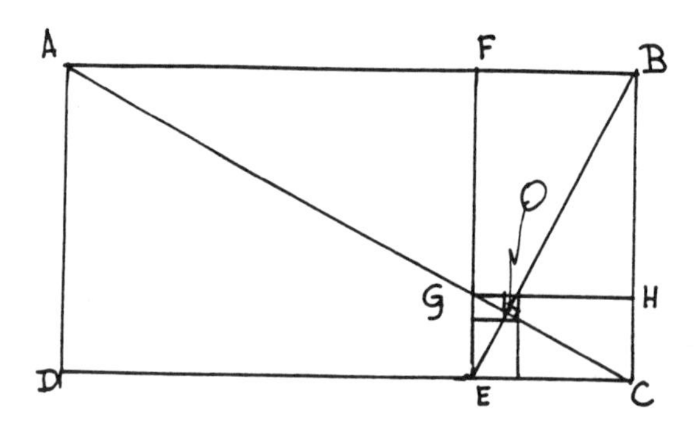

Figure 19-2 ***Continuation of the division process***

Q: *COULD YOU FIND A RECTANGLE WHICH BY AN ANALOGOUS PROCESS WOULD GENERATE RECTANGLE ABCD?*

The original rectangle ABCD may be seen to be itself generated from a similar rectangle by extending the line BE till it intersects the extension of AD (Figure 19-3). Through the point of intersection J draw a line parallel to AB, which intersects the extension of BC at K. Rectangle JABK is then similar to rectangle ABCD, its gnomon is JDCK.

All rectangles similar to ABCD in Figure 19-2 have in common the point of intersection of their diagonal with the line drawn perpendicular to it; call this point O. Then **OB/OA = OC/OB = OE/OC**, etc. Since the triangle AOB is similar to triangle ABC, this ratio is also the ratio of the short to the long sides of every rectangle similar to ABCD. Let us label this ratio f. We can then imagine the sequence of similar rectangles to be generated by a rotation of ninety degrees around O, accompanied by a reduction in scale by factor f.

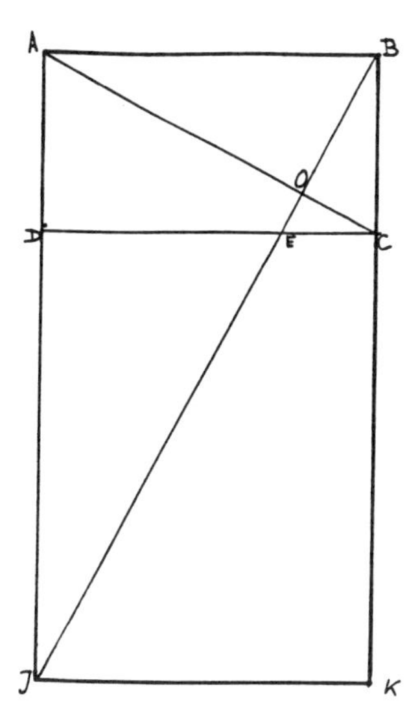

Figure 19-3 ***Finding the rectangle which on analogous division would produce the original rectangle of Figure 19-1***

The relation between successive rectangles so constructed is called *dynamic symmetry*[1]. Ordinarily the term *symmetry* is reserved for the relationship between mutually congruent objects; the qualification *dynamic* apparently was introduced to relax the condition of *congruency* to the less rigorous one of *similarity*. We note that the scale of each rectangle is reduced by the same factor with each ninety degree rotation: The relative rate of reduction remains constant. Accordingly, the points A, B, C, E, G, etc., lie on a common logarithmic spiral. Such a spiral has the same function in dynamic symmetry as a circle does in rotational symmetry. Since all dimensions of the rectangles are correspondingly reduced, the points D, F, H, etc., will also lie on a common logarithmic spiral, albeit different from the first spiral.

Q: *HOW WOULD YOU DRAW A LOGARITHMIC SPIRAL THROUGH THESE POINTS ON YOUR RECTANGLE?*

To draw a logarithmic spiral through these points it is necessary to interpolate distances from O for all angles measured around O, such that the spiral passes through the desired points at intervals of 90°. This is easily done once it is recalled that for a logarithmic spiral the graph on semi-log paper of the radial coordinate versus the angular coordinate is a straight line. Therefore only two points are needed to plot the logarithm of the radial coordinate against the angular coordinate. Intermediate values can then be read off the semilog graph as needed to produce a curve that appears smooth within the desired level of resolution and plotted on polar graph paper to produce the desired logarithmic spiral (cf. Appendix 2).

We have now generated logarithmic spirals by two entirely different processes: The dynamic symmetry construction and the bugs' walk. The former method is valid for any value of the scaling factor f, whereas the latter produced a very special spiral, namely one which always makes a 45° angle with the radial direction. Recalling that the equation of the bugs' spiral is $r = r_0 e^{-\theta}$, we find that over 90° this function is reduced by a factor of 0.21, which thus corresponds to the factor f for the bugs' spiral.

Q: *TRACE THE BUGS' SPIRAL, AND SUPERIMPOSE ON IT THE RECTANGLE CORRESPONDING TO THIS SPIRAL.*

In the present chapter we shall consider yet another important particular configuration, namely the rectangle whose gnomon is a square. Since this is a very unique rectangle, we use a special symbol for the ratio of the length of the shorter side to that of the longer one, namely ϕ. If we assign to **AB** a unit of length (cf. Figure 19-2), then **BC**$= \phi$. Since **CE/BC** also equals ϕ, **CE**$= \phi^2$. Since **DC**$=1$, **DE**$= 1 - \phi^2$. As a result, the condition that the gnomon rectangle AFED be a square is

$$1 - \phi^2 = \phi. \qquad (19-1)$$

It will be of interest to find out whether a rectangle having a square gnomon can be plotted on quadrille paper such that all four of its vertices lie at the corners of the quadrille paper's squares. To fit a rectangle onto quadrille paper like this, ϕ would need to be a rational fraction. If it is, then it could be expressed as a ratio of integers which have no common factor. (If they did have common factors, these would be factored out.) Therefore, if we write $\phi \equiv v/v$, where u is the number of squares along the short side of the rectangle and v the number along the long side, then equation (19-1) becomes

$$v^2 - u^2 = uv.$$

Factoring the left hand side:

$$(v + u)(v - u) = uv. \qquad (19-2)$$

Both u and v may be either odd or even. Remembering that the sum and difference of two odd or two even numbers are always even, that the sum and difference of an odd and an even number are always odd, that the product of an even integer with any other integer is even, and that the product of two odd numbers is odd, we find the following possibilities.

Table 19-1: ***Parities of the Sides of Equation 19-2.***

If u is:	and v is:	Then $(v+u)$ is:	$(v-u)$ is:	$(v+u)(v-u)$ is:	uv is:
even	even	even	even	even	even
even	odd	odd	odd	odd	even
odd	even	odd	odd	odd	even
odd	odd	even	even	even	odd

This table tells us that, since an odd integer cannot equal an even one, both sides of equation (19-2) can be equal to each other only if both u and v are even. However, we must discount this possibility because u and v were supposed not to have a common factor, and two even numbers have the factor 2 in common. We conclude that the assumption that ϕ might be rational leads to an absurdity, and must be discarded.

Of course, we could have "solved" the quadratic equation (9-1), finding, since ϕ must be positive, $\phi = \frac{1}{2}(\sqrt{5} - 1)$. This however, is a cop-out, for how much more do we know about the expression $\sqrt{5}$ than about the number ϕ? As before, we shall see that equation which defines the quantity tells us more than its so-called solution. Before we examine some further remarkable properties of ϕ, we should meditate briefly on the significance of its being irrational, and the way in which we convinced ourselves that it is.

The rectangle whose gnomon is a square is called the *golden rectangle*, and the fraction ϕ the *golden fraction*. The proportions of these golden figures have been since classical antiquity considered agreeable, and even magical.[2] There have been attempts to construct a golden rectangle by putting squares together, and intuitively it is not at all obvious that this would not be possible. Significantly, the argument which we used above to prove this impossibility did not involve any complex mathematical manipulations, but simply logical reasoning.

We noted that successive rectangles constructed within a golden rectangle by the dynamic-symmetry method will have their short sides equal to successive powers of ϕ. Since ϕ is irrational, it has infinitely many digits behind the decimal point; neglecting all but a finite number will eventually cause large round-off errors when high powers are computed. It is comforting to know, therefore, that we never need higher powers of ϕ than the first, because we can express all its higher powers in terms of the first as follows. If we multiply both sides of equation (19-1) by ϕ, $\phi^3 = \phi - \phi^2$. Substituting for ϕ^2 the expression given by equation (19-1), we find that $\phi^3 = 2\phi - 1$.

Q: *FIND HIGHER POWERS OF ϕ EXPRESSED IN TERMS OF THE FIRST POWER OF ϕ.*

Analogously, we can reduce the higher powers of ϕ to lower ones, generating the following table of powers.

Table 19-2: ***Powers of ϕ.***

$$\phi = 1\phi - 0$$
$$\phi^2 = 1 - 1\phi$$
$$\phi^3 = 2\phi - 1$$
$$\phi^4 = 2 - 3\phi$$
$$\phi^5 = 5\phi - 3$$
$$\phi^6 = 5 - 8\phi$$
$$\phi^7 = 13\phi - 8$$
$$\phi^8 = 13 - 21\phi$$
$$\phi^9 = 34\phi - 21$$

etc.

If we recall that ϕ is positive, then we realize that the right hand sides of these equations must also be positive, so that we learn from the successive equations:

Table 19-3: ***Upper and Lower Bounds on ϕ.***

$\phi > 0$	$0/1 = 0.000$
$\phi < 1$	$1/1 = 1.000$
$\phi > 1/2$	$1/2 = 0.500$
$\phi < 2/3$	$2/3 = 0.667$
$\phi > 3/5$	$3/5 = 0.600$
$\phi < 5/8$	$5/8 = 0.625$
$\phi > 8/13$	$8/13 = 0.615$
$\phi < 13/21$	$13/21 = 0.619$
$\phi > 21/34$	$21/34 = 0.618$

Thus the irrational number ϕ is successively bounded from below and above by simple rational fractions, so that we may approximate it to within any desired level of resolution by a rational number; to three decimal places it may be approximated by 0.618.

These bounding fractions are nicely visualized by the following construction. Start with a small square, and put another square adjacent to it, sharing an edge with it. The result is, of course, a 1-by-2 rectangle. On the long side, add on a square, flush with the rectangle. There results a 2-by-3 rectangle (cf. Figure 19-4). At each stage we have a square adjacent to a rectangle; unlike the dynamic-symmetry construction, the rectangles are not mutually similar, but their proportions are successively the bounds on ϕ listed in Table 19-3. Therefore this construction will gradually approach the golden rectangle, but will never quite achieve it, as it will always produce a rational rectangle, i.e., one whose sides are multiple lengths of a common divisor.

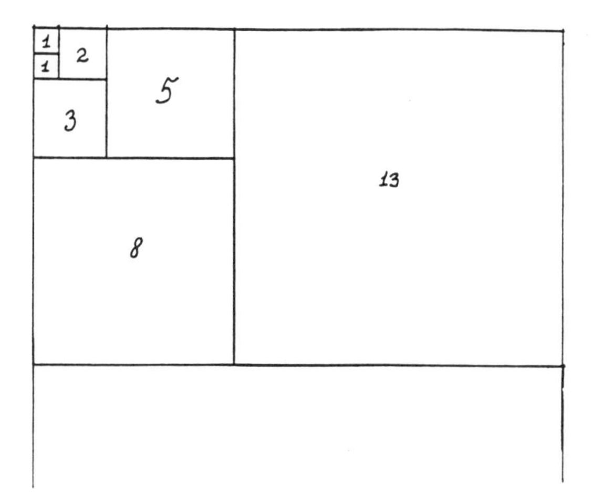

Figure 19-4 ***Graphic representation of Fibonacci numbers***

The sides of successive rectangles just constructed are the integers 1, 1, 2, 3, 5, 8, etc. These numbers are generated by having each be the sum of the previous two. They are known as the *Fibonacci numbers*.[3] Table 19-3 tells us that the ratio between successive Fibonacci numbers more closely approaches the golden fraction the further into the series we are. We may define the Fibonacci series recursively: If we denote two successive Fibonacci numbers respectively by a_{n-1} and a_n, then the next number, a_{n+1} is given by

$$a_{n+1} = a_n + a_{n-1} \,. \qquad (19-3)$$

If we divide both sides of equation (19-3) by a_{n+1}, and multiply and divide the second term on the right by a_n in order to use ratios of adjacent Fibonacci numbers only:

$$1 = a_n/a_{n+1} + (a_{n-1}/a_n)(a_n/a_{n+1}) \,.$$

We observed that the ratio of successive Fibonacci numbers appears to approach a constant value: If we choose n sufficiently large, then within our level of resolution we may set this ratio equal to a constant, which we shall call K. Substitution in the last equation gives

$$1 = K + K^2 .$$

This equation is equivalent to equation (19-1), so that we have demonstrated again that, if the ratio of successive Fibonacci numbers approaches a constant, that constant must indeed be the golden fraction.

This result was derived from the recursion formula (19-3); no use was made of what the initial pair of Fibonacci numbers is.

Q: *THE IMPLICATION IS THAT ONE COULD START WITH ANY PAIR OF POSITIVE INTEGERS, AND, USING THE SAME GENERATING METHOD AS USED FOR THE FIBONACCI SERIES, PRODUCE A SERIES IN WHICH THE RATIO OF SUCCESSIVE NUMBERS WOULD APPROACH THE GOLDEN FRACTION! DOES THAT APPEAR REASONABLE? CHOOSE AN ARBITRARY PAIR, AND CHECK WHAT THE RATIO WOULD EVENTUALLY BE.*

It is logical, if not intuitively obvious, to conclude that starting with any pair of positive integers one could use the same recursive processes to generate a sequence of numbers whose ratios will ultimately approach the golden fraction. Below are a few examples to test this conclusions:

1 4 5 9 14 23 37 60 97 157 254; last ratio = 0.618
93 45 138 183 321 504 825 1329; last ratio = 0.621
1 99 100 199 299 498 797; last ratio = 0.625.

In the last two sequences the numbers eventually come close to one hundred times the original Fibonacci numbers, so that their successive ratios will certainly approach those of the original sequence.

We conclude, therefore, that it is the recursive relationship, not the particular choice of the initial pair of numbers, which causes the ratio of successive numbers to approach the golden fraction, ϕ. It is this property which gives ϕ such a universal meaning. Like e, it is irrational, but like e, it can be approached within any desired level of resolution by a rational number.

There is also a recursive expression for ϕ. From equation (19-1) we find

$$\phi = 1/(1+\phi) . \qquad (19-4)$$

We substitute in the right hand side of this last equation the expression given for ϕ by the equation $\phi = 1/(1+1/(1+\phi))$. Repeating this substitution,

$$\phi = 1/(1+1/(1+1/(1+1/(1+\phi)))), \quad \text{etc.,}$$

so that

$$\phi = 1/(1+1/(1+1/(1+1/(1+1/(1+\ldots)\ldots)))) .$$

This expression is useful when ϕ needs to be calculated with very high accuracy on a computer.

Finally, a useful and interesting relation between any power of ϕ and two successive Fibonacci numbers can be derived. Table 19-2 may be generalized in the form

$$\phi^n = (-1)^{n+1}(p_n\phi - q_n) , \qquad (19-5)$$

where p_n and q_n are constants still to be determined.

To determine these two constants, we obtain ϕ^{n+1} from equation (19-5) in two different manners, and then equate the results. First we rewrite equation (19-5) substituting $n+1$ for n:

$$\phi^{n+1} = (-1)^{n+2}(p_{n+1}\phi - q_{n+1}) . \qquad (19-6)$$

Secondly, we multiply both sides of equation (19-5) by ϕ:

$$\phi^{n+1} = (-1)^{n+1}(p_n\phi^2 - q_n\phi) .$$

Since $\phi^2 = 1-\phi$, this last equation becomes

$$\phi^{n+1} = (-1)^{n+2}[(p_n + q_n)\phi - p_n] . \qquad (19-7)$$

Equations (19-6) and (19-7) are equivalent if $q_{n+1} = p_n$, and $p_{n+1} = p_n + q_n$, so that $p_{n+1} = p_n + p_{n-1}$. This is precisely the recursion relation found for the Fibonacci numbers a_n; since from Table 19-2 we know that $p_1 = 1$ and $p_0 = 0$, we may conclude that the numbers p_n are in point of fact the Fibonacci numbers, so that

$$\phi^n = (-1)^{n+1}(a_n\phi - a_{n-1}) . \qquad (19-8)$$

Since $\phi < 1$, ϕ^n will tend to zero when n becomes very large. As a result the right hand side of equation (19-8) will also tend to zero when n becomes very large, so that we are once more assured that when n is sufficiently large, ϕ may be closely approximated by the ratio of a_{n-1} to a_n. Moreover, the left hand side of that equation yields an estimate of the magnitude of the error implied by this approximation.

Equation (19-4) tells us that the reciprocal of the golden fraction equals the golden fraction plus unity. That reciprocal, $\tau \equiv 1+\phi \equiv 1/\phi$, is often used as the golden ratio. We feel that the interval between zero and unity is a rather unique one, and therefore defined the golden fraction ϕ to lie in that range. For instance, we have made extensive use of the fact that powers of ϕ tend to zero. However, the number τ also has remarkable properties, just as ϕ does:

$$\tau = 1+\phi \qquad 1/\tau = \phi .$$

Subtract these expressions: $\tau - 1/\tau = 1$, hence $\tau^2 - \tau = 1$.

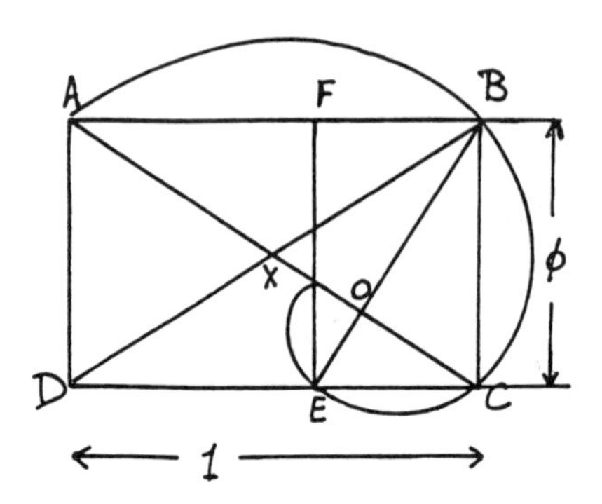

Figure 19-5 ***Golden rectangle***

If ABCD is a golden rectangle, then the tangent of angle CAB equals ϕ.[4] Figure 19-5 shows a golden rectangle ABCD, the line EF dividing ABCD into the rectangle BCEF similar to ABCD, the square gnomon AFED, both diagonals of ABCD, AC and BD which intersect each other at X, the spiral proper to ABCD, and O, the center of that spiral. Angle BXC equals twice angle BAC:

$$\text{Angle}\quad \text{BXC} = 2 \arctan \phi \,.$$

Using the formula $\tan 2\alpha = 2 \tan \alpha / (1 - \tan^2 \alpha)$:

$$\tan\quad \text{BXC} = 2\phi/(1-\phi^2)\,.$$

From equation (19-1) we know that $1-\phi^2 = \phi$, so that $\tan\text{BXC} = 2$. Therefore the length of OB is exactly twice that of OX. This result leads to the following conclusions:

Theorem 19-1: The diagonals of each golden rectangle intersect at an angle whose tangent equals 2. The center of a spiral proper to a golden rectangle lying on one of its diagonals is exactly twice as far from the nearer end of the other diagonal as from the intersection of both diagonals.

Theorem 19-2: The golden fraction is given by the expression

$$\phi = \tan(\frac{1}{2} \arctan 2)\,. \qquad (19-9)$$

Q: *USE EQUATION (19-9) TO CONSTRUCT A GOLDEN RECTANGLE AS SIMPLY AS YOU CAN.*

Equation (19-9) gives an explicit expression for the golden fraction. Admittedly it uses transcendental trigonometric functions which may not be considered superior to the use of $\sqrt{5}$. However, equation (19-9) allows a simple construction of a golden rectangle: The diagonals are simply the sides of an angle whose tangent equals 2, and hence is easily drawn (Figure 19-6):

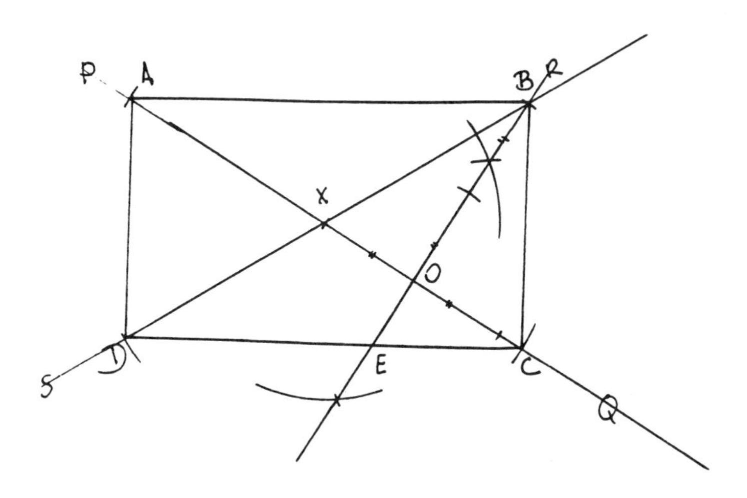

Figure 19-6 ***Construction of a golden rectangle***

Step 1 : Draw one diagonal (PQ).
Step 2 : Draw a perpendicular to the diagonal (OR).
Step 3 : Choose X on PQ.
Step 4 : Measure off twice the distance OX along OR to locate B.
Step 5 : Draw a second diagonal (XB).
Step 6 : Locate vertices A, C and D on the diagonals at the same distances from X as was B.
Step 7 : Draw AB, BC, CD and DA.

If the length of the diagonals of the desired rectangle is known, then the vertices of this rectangle are simply the intersection of the diagonals with the circle centered on the intersection of these diagonals having as diameter the desired length of the diagonals. On the other hand, if the desired golden rectangle is to have a given side length, then the angle between the diagonals is easily bisected, and the desired length measured off along the appropriate bisector.

NOTES

[1] Edwards, Edward B.: *Patterns and Designs with Dynamic Symmetry,* Dover, New York (1967).

[2] Loeb, A. L.: *The Magic of the Pentangle: Dynamic Symmetry from Merlin to Penrose*, J. Computers & Mathematics with Applications, **17**, 33–48 (1989), and *Symmetry 2, Unifying Human Understanding*, I. Hargittai, ed., Pergamon, New York/Oxford (1989).

[3] This series is named for Leonardo Fibonacci, also known as Leonardo da Pisa (ca. 1170–after 1240), an Italian mathematician who advocated the use of Arabic notation for numerals.

[4] The remainder of this chapter is adapted from Loeb, Arthur L. and William Varney: *Does the Golden Spiral Exist, and If Not, Where Is Its Center?* In: *Spiral Symmetry,* I. Hargittai ed., World Science Books, Singapore, 47–61 (1991).

XX

The Golden Triangle

Triangle ABC (cf. Figure 20-1) is isosceles: **AB = AC**. A is called its apex, BC its base. The apex angle is less than sixty degrees. Another triangle, BCD, is constructed inside ABC, which is also isosceles and has B as its apex; the point D lies on AC. Since their base angles are the same, triangles ABC and BCD are similar. In analogy to dynamic symmetry nomenclature we shall call triangle DAB the gnomon of ABC.

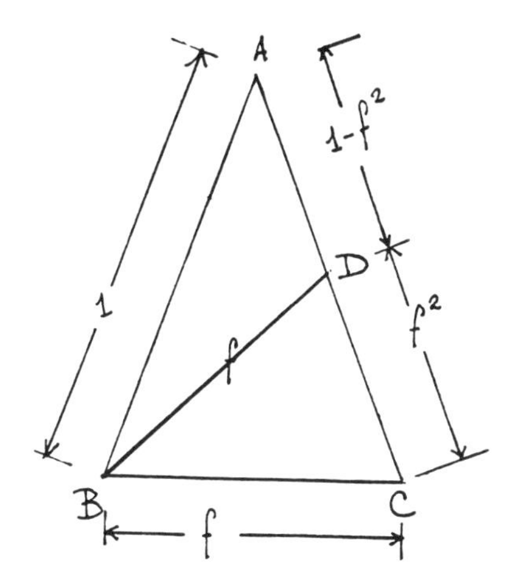

Figure 20-1 ***An isosceles triangle***

Let us assign unity to the length **AB**. Let us denote the base **BC** by f. Since the angle A is less than sixty degrees, f is less than unity. Because of the similarity of the two triangles ABC and BCD, **CD**$= f^2$. Therefore **AD**$= 1 - f^2$. Because triangle BCD is isosceles, **BD**$= f$.

Q: *COULD THE GNOMON ALSO BE ISOSCELES?*

The gnomon DAB is isosceles if **DA** = **DB**, i.e.,

$$f = 1 - f^2 .$$

This equation is recognized as equation 19-1, the definition of the golden fraction ϕ. An isosceles triangle whose ratio of baselength to sidelength is the golden fraction is called a *golden triangle*. We shall call its gnomon the *golden gnomon* (Figure 20-2).

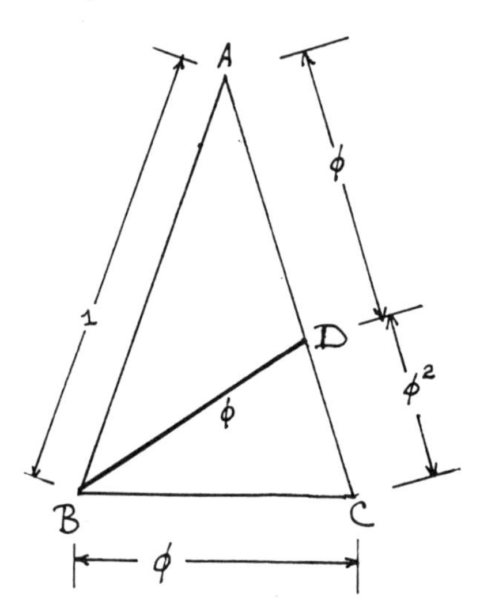

Figure 20-2 ***A golden triangle subdivided into a smaller golden triangle and a golden gnomon***

The distances **AD** and **BD** are, accordingly, both equal to ϕ, so that the golden triangle has a ratio of base length to side length equal to the golden section ϕ, whereas the golden gnomon has the ratio of side length to base length equal to the golden section ϕ. We may add to the golden triangle ABC a golden gnomon placed with its apex at B, and having C as a terminal of its base (Figure 20-3). Call the other base terminal E. The distance **AE** equals the distance **AB** plus the distance **BE**, i.e., $1+\phi$. Recall that $1+\phi = 1/\phi(\equiv \tau)$, so that the triangle CAE has its side to base lengths in the golden ratio, hence is a golden gnomon. Figure 20-2 shows that a golden triangle and a golden gnomon juxtaposed so that they share a side may generate a new golden triangle whose dimensions are $1/\phi$ times as large as the first golden triangle. Figure 20-3 shows that a golden gnomon added to a golden triangle such that a side of the gnomon coincides with the base of the golden triangle generates a golden gnomon whose dimensions also are in ratio $1/\phi(\equiv \tau)$ to those of the original gnomon.

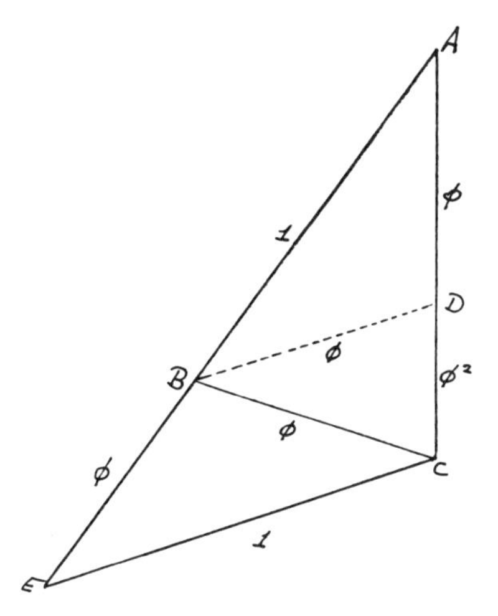

Figure 20-3 ***Golden triangle together with golden gnomon***

Q: *CREATE TILINGS USING BOTH THE GOLDEN TRIANGLE AND THE GOLDEN GNOMON. NOTE THAT HERETOFORE WE CONSIDERED TESSELLATIONS IN WHICH ALL TRIANGLES ARE MUTUALLY CONGRUENT. WOULD THE GOLDEN TRIANGLE AND THE GOLDEN GNOMON TILE THE PLANE INDIVIDUALLY?*

Golden triangles paired with golden gnomons can therefore be used as modules to create larger golden triangles and gnomons scaled up by the golden section; this process may be continued indefinitely. The angles of these triangles are found as follows from Figure 20-1.

Call angle BAC= α. Then angle ACB= $\frac{1}{2}(180° - \alpha)$ =angle BDC. In triangle DAB, angle ADB= $180° - 2\alpha$. Since angles ADB and BDC add up to 180°, $\frac{1}{2}(180° - \alpha) + (180° - 2\alpha) = 180°$, hence $\alpha = 180°/5 = 36°$.

Accordingly, the angles of the golden triangle are $36°, 72°$ and $72°$, and those of the golden gnomon are $108°, 36°$ and $36°$. All these angles are integer multiples of 36°, another reason why the golden triangle and the golden gnomon are good modular building blocks.

For example, in Figure 20-4 we have placed two golden gnomons with their bases coinciding with the sides of unit length of a golden triangle: The result is seen to be a regular pentagon whose sides are of length ϕ, and whose angles are 108°. We conclude that the regular pentagon has sides and diagonals whose respective lengths are in the ratio of the golden fraction. Figure 20-5 is a regular pentagon "stellated" by five golden triangles: The result is a five-pointed star or *pentagram*. If these golden triangles have a base length equal to ϕ, their sides, and hence the sides of a pentagram, have

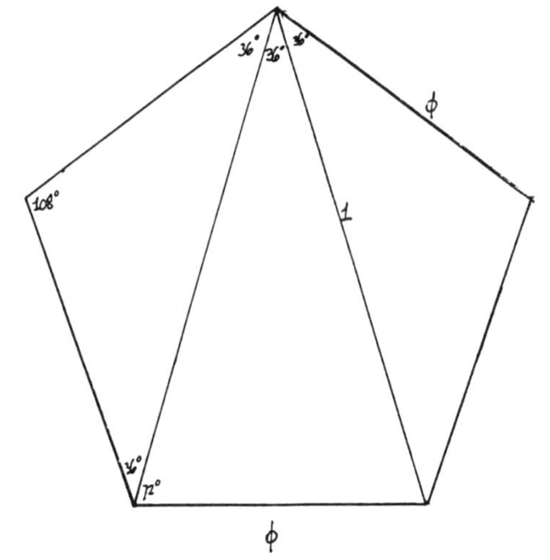

Figure 20-4 ***Pentagon generated by a golden triangle together with two golden gnomons***

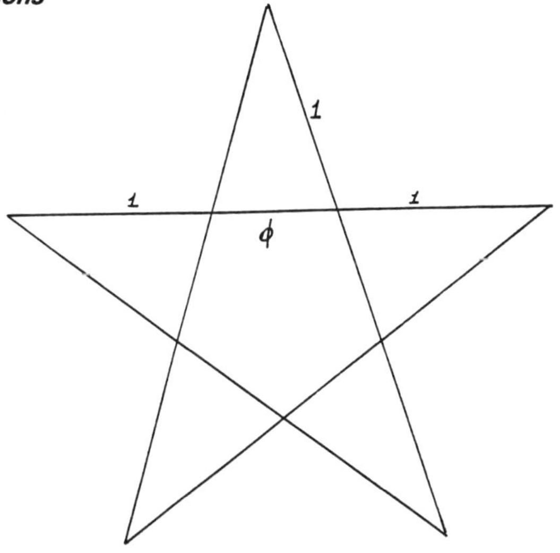

Figure 20-5 ***Regular pentagram***

unit length. In Figure 20-6 the points of the star are joined: The spaces between the points of the star are filled in by golden gnomons. The result is a regular pentagon having edgelength $1 + \phi(\equiv \tau)$, which was shown to be the same as $1/\phi$. Since the original pentagon has edgelength ϕ, the new and the old pentagons have their edgelengths in the ratio

$$\frac{(1+\phi)}{\phi} = \frac{1}{\phi} + 1 = (1+\phi) + 1 = 2 + \phi \equiv 1 + \tau$$

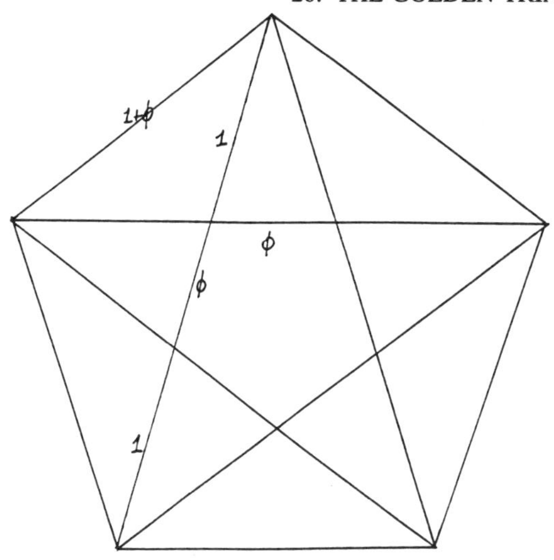

Figure 20-6 ***Pentagram and pentagon***

Figure 20-7 shows yet another construction of a regular pentagon out of golden triangles and golden gnomons: Two gnomons placed base-to-base, forming a rhombus having angles of 108° and 72° and sides of length ϕ. Two golden triangles having side lengths equal to ϕ are added; their base length equals ϕ^2. Finally, a golden gnomon having side length equal to ϕ^2 fills in the remaining space to complete the regular pentagon having edgelength equal to ϕ.

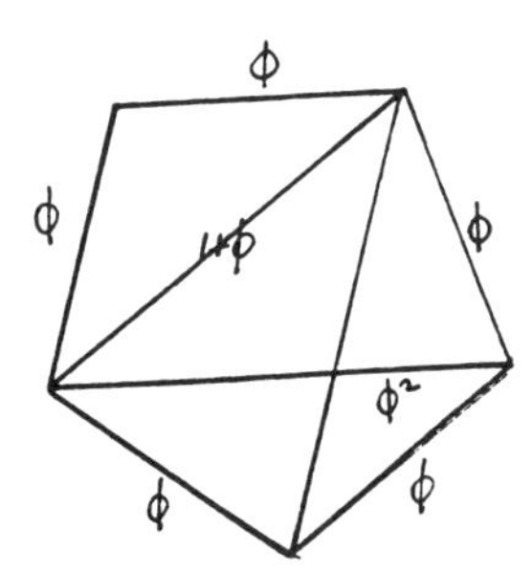

Figure 20-7 ***Regular pentagon generated by golden triangles and golden gnomons***

Evidently we could cover any surface with combinations of golden triangles and golden gnomons without leaving any gaps or having any tiles overlap, for we have demonstrated that we can scale any of them up by the golden ratio, hence repeat the processes just described *ad infinitum*. However, in Theorem 5-3 we noted that a five-fold rotocenter in a plane precludes any other rotocenter in that plane. Thus tessellations using golden triangles and gnomons together may have at most a single five-fold rotocenter. Tessellations using golden triangles and gnomons were studied by Roger Penrose,[1] and discussed by Martin Gardner.[2]

Q: *DIVIDE TRIANGLE BDC INTO A GOLDEN TRIANGLE AND A GOLDEN GNOMON, AND CONTINUE THE PROCESS FOR EACH SUCCESSIVE GOLDEN TRIANGLE GENERATED. NOTE THAT IT MAKES A DIFFERENCE WHERE YOU PLACE THE APEX OF THE NEW GOLDEN TRIANGLE.*

Triangle BDC, a golden triangle, may be in turn subdivided into a golden triangle and a golden gnomon, and the process continued (Figure 20-8). This figure exhibits dynamic symmetry[3]: In this instance a rotation of 108° is accompanied by a reduction in size by a factor of ϕ. A logarithmic spiral can be drawn through successive dynamically related points (Figure 20-9), but whereas in the case of the golden *rectangle* a *90°* rotation is accompanied by a scaling factor of ϕ, for the golden *triangle* the same scaling occurs over a rotation by *108°*. It is therefore not possible to apply the name *golden* spiral uniquely to *both* the logarithmic spiral proper to the golden rectangle and to that proper to the golden triangle.

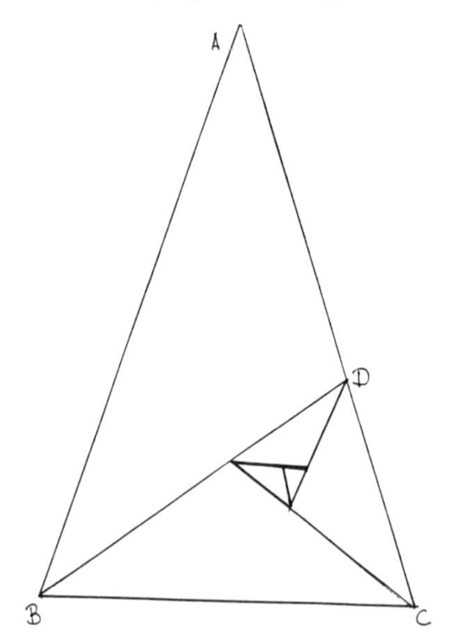

Figure 20-8 ***Continuation of the subdivision of the golden triangle***

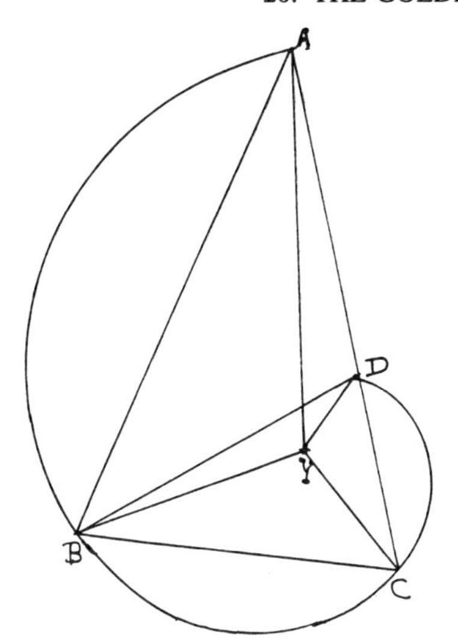

Figure 20-9 ***Golden triangle and logarithmic spiral***

In contrast to the case of the spiral proper to the golden rectangle (cf. Chapter XIX), it is not at all obvious where the center of the spiral proper to the golden triangle is located. Huntley[4] states that it lies on the median line joining a base vertex to the center of the opposite side, but does not provide a proof.

Q: *USE HUNTLEY'S CONJECTURE TO LOCATE THE CENTER OF THE SPIRAL PROPER TO THIS DYNAMICALLY SYMMETRICAL PATTERN.*

If Huntley's assertion is indeed correct, then corresponding medians in nested golden triangles will intersect at the center of the spiral, as this center necessarily lies on the corresponding median of each of these nested golden triangles. We shall return to this assertion presently.

It will be convenient, though, to introduce first a *golden parallelogram*, and then to find the center of the logarithmic spiral proper to the golden triangle. Figure 20-10 shows a parallelogram PQRS having angles 108° and 72° and sides whose lengths are in the golden ratio. As we shall be considering this parallelogram in various orientations and scalings, we shall give the lengths of its sides PS and PQ the general designations a and $a\phi$ respectively.

From the law of cosines we calculate the diagonals of this golden parallelogram:

$$\begin{aligned} \mathbf{PR}^2 &= a^2(1+\phi^2) - 2a^2\phi\cos 72^\circ \\ \mathbf{QS}^2 &= a^2(1+\phi^2) + 2a^2\phi\cos 72^\circ\,. \end{aligned}$$

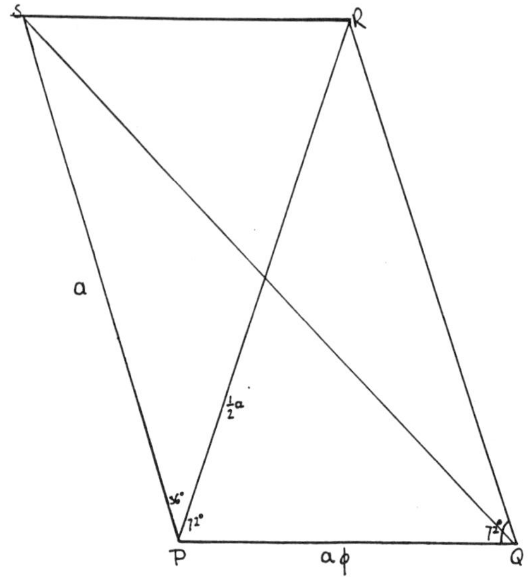

Figure 20-10 ***A golden parallelogram***

Q: *USING THE DIMENSIONS OF THE GOLDEN TRIANGLE, FIND THE RELATION BETWEEN* $\cos 72°$ *AND THE GOLDEN FRACTION.*

As 72° is the value of the base angles of a golden triangle, and the ratio of base length to side length of the golden triangle is ϕ, $\cos 72° = \frac{1}{2}\phi$, so that

$$\mathbf{PR} = a$$

$$\mathbf{QS} = a\sqrt{1 + 2\phi^2}\,.$$

Accordingly, the golden parallelogram is divided by its short diagonal into two golden triangles. As two diagonals of a parallelogram bisect each other, the long diagonal is a median line in each of these golden triangles.

The long diagonal divides the golden parallelogram into two mutually congruent triangles which are not as familiar as are the golden triangles, but which are nevertheless quite remarkable, and will be helpful in locating the center of the spiral proper to the golden triangle. Because of their special relationship to the golden triangle we shall call them golden supplements. Triangle QPS has an angle of 108°; the sides of this angle have lengths related by the golden ratio. The short diagonal of the golden parallelogram is a median line of the golden supplement. This median line has length $\frac{1}{2}a$ and divides the 108° angle into one angle of 72° and one of 36°. Hence:

Theorem 20-1: The median line joining the vertex at the 108° angle of the golden supplement to the center of its opposite side has a length exactly half that of that obtuse angle's longer side, and divides that obtuse angle into two unequal portions of which the smaller is exactly one half as large as the latter. The side opposite the 108° angle has length $\sqrt{1+2\phi^2}$ times that angle's longer side.

Figure 20-11 shows the golden triangle ABC, the line AD subdividing it into a golden triangle BCD and a golden gnomon ADB, and a point Y, the center of the spiral proper to these triangles, whose location is to be determined. Since Y is a center of dynamic symmetry, and the points A and B are respective apices of triangles related by dynamic symmetry, angle AYB= 108°, and **YB/ YA=** ϕ. Accordingly, triangle AYB is a golden supplement. By the same reasoning it is concluded that the triangle BYC is a golden supplement.

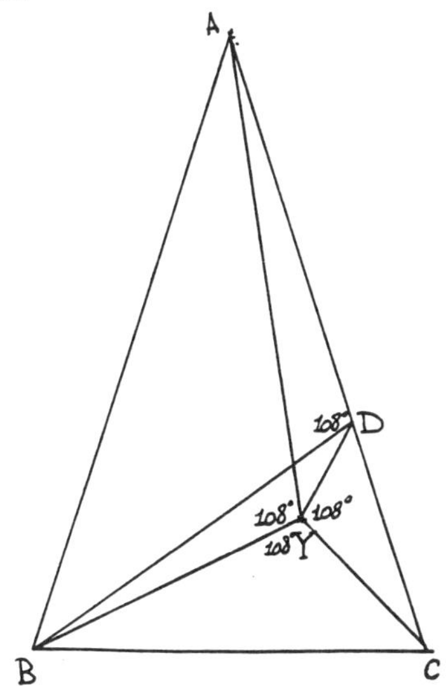

Figure 20-11 ***Center of dynamic symmetry in a set of nested golden triangles***

The point Y subtends a circular arc of 108° at the line AB. The same is true of the point D. Therefore points A, Y, D and B lie on a common circle (Figure 20-12), the circle circumscribed around the golden gnomon ADB. The same reasoning applies to the entire set of nested golden gnomons, and leads to Theorem 20-2.

Theorem 20-2: In a set of nested golden triangles and golden gnomons related by dynamic symmetry, all circles circumscribing the golden gnomons pass through a common point which is the center of the logarithmic spiral proper to the nested golden triangles and gnomons.

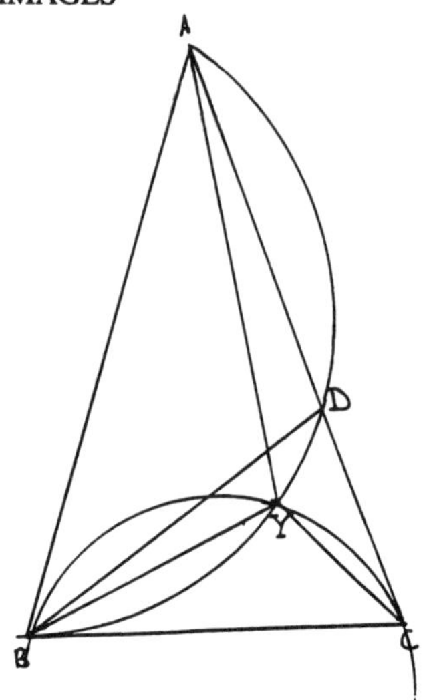

Figure 20-12 ***Center of dynamic symmetry lies on circles circumscribed around the golden gnomons***

Theorem 20-2 provides a simple construction method for locating the center of our spiral. It provided the inspiration for artist Damian Bagdan's design (Figure 20-13). We have, however, not yet fully exploited the properties of the golden supplement. To do so, extend line CY to meet AB at

Figure 20-13 ***Design by Damian Bagdan*** (From the Teaching Collection of the Carpenter Center for the Visual Arts at Harvard University. Reproduced with permission of the curator. Photography by C. Todd Stuart.)

point Z (Figure 20-14). Since angles CYB and BYA both equal 108°, angle AYC equals 144°. Therefore angle ZYA equals 36°, with the result that angle ZYB equals 72°. Since we have seen that triangle AYC is a golden supplement, the line YZ, which divides the angle AYC into portions of 72° and 36°, is a median of triangle AYC. Therefore **AZ** = **BZ**, so that line BZ is a median line of triangle ABC, and Huntley's assertion is proven correct. Since the side AB of the golden triangle ABC was set at unity, we find from Theorem 20-1 that the sides of the golden supplement AYB have lengths **AY**$= 1/\sqrt{1+2\phi^2}$, **BY**$= \phi/\sqrt{1+2\phi^2}$ and **AB**$= 1$, and that the length of its median line YZ equals $1/2\sqrt{1+2\phi^2}$.

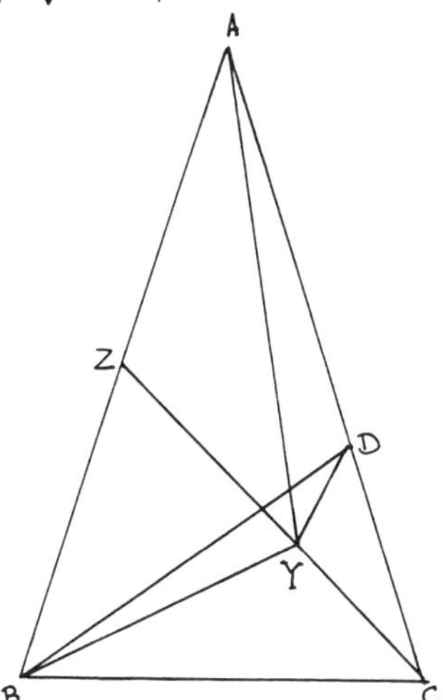

Figure 20-14 ***Proof of Huntley's assertion***

These results may be quite lucidly summarized in terms of a golden parallelogram ACBF constructed so that triangle ABC is the original golden triangle, and F is the intersection of a line drawn through the point A parallel to BC with a line drawn through the point B parallel to AC (Figure 20-15). Triangle BAF is, of course, golden. The long diagonal CF passes through the centers of two logarithmic spirals, one, Y, proper to golden triangle ABC, the other, X, proper to the golden triangle BAF. Either of these spirals may be called proper to the golden parallelogram ACBF, and from the values found above for distances **AY, BY** and **YZ** we derive the following theorem:

Theorem 20-3: The center of a logarithmic spiral proper to a golden parallelogram lies on the longer diagonal of that parallelogram, exactly twice as close to the intersection of the two diagonals as to the farther end of the shorter diagonal.

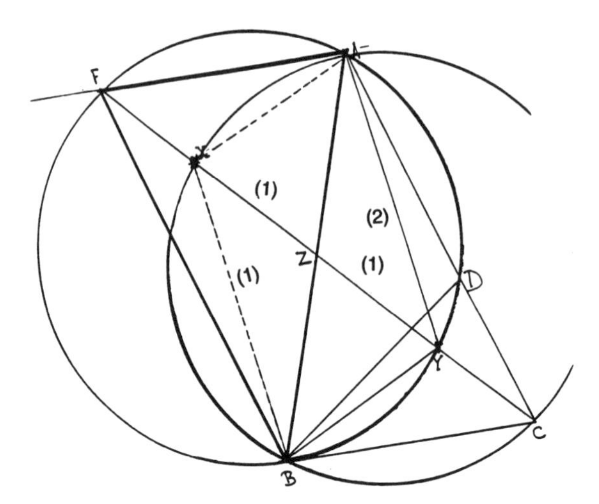

Figure 20-15 ***Golden parallelogram together with the centers of its proper spirals***

Q: *WHAT ANGLE DOES POINT F SUBTEND AT LINE AB? FIND ITS LOCATION WITH RESPECT TO THE CIRCLE PASSING THROUGH A, D, Y AND B.*

We noted that the spiral center Y, subtending an angle of 108° at the points A and C, lies on the same circle which circumscribes the golden gnomon ABD. The point F subtends an angle of 72° at points A and B, and therefore also lies on the circle passing through points A, D, Y and B. Thus:

Theorem 20-4: The twin golden triangles constituting a golden parallelogram are paired so that the spiral center proper to one lies on the circle circumscribing the other.

The last observation leads to a simple construction method for locating the center of the logarithmic spiral proper to a golden triangle. Remember that the center of the circle circumscribing a triangle is at the intersection of the perpendicular bisectors of the sides of that triangle. If we consider golden triangle ABC, and recall that the circle circumscribing its gnomon ABD will also circumscribe the twin golden triangle BAF, then we conclude that the center of that circumscribing circle lies at the intersection of the perpendicular bisector of AB with the perpendicular to BC drawn through B, the latter being also the perpendicular bisector of AF. Thus we have used the concept of the golden parallelogram without actually needing to draw it explicitly in the construction process.

The construction method accordingly is as follws (cf. Figure 20-16).

Step 1 : Draw the perpendicular bisector of AB, calling its intersection with AB Z. The intersection of this perpendicular bisector with AC will be D, as, by definition of the golden triangle, the gnomon ADB is isosceles.

Step 2 : Draw CZ.

Step 3 : Draw a perpendicular to BC through B.

Step 4 : Call the intersection of this latter perpendicular with the perpendicular bisector of AB U.

Step 5 : With U as the center and UB as a radius, draw a circle. The intersection of this circle with CZ will be the desired center, Y. If the drawing has been accomplished accurately, the last circle will also pass through point D.

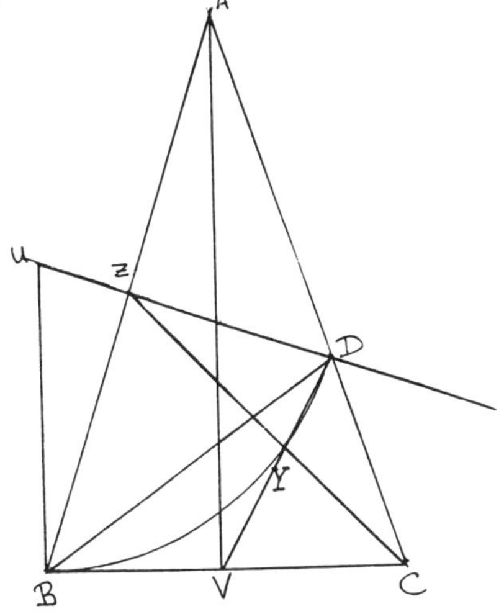

Figure 20-16 ***Construction of the center of the logarithmic spiral proper to dynamically nested golden triangles***

An alternate construction would involve dropping a perpendicular from A onto BC rather than erecting a perpendicular to BC at A. This alternative perpendicular would intersect the base BC at its center V, which would in turn be connected to point D by a straight line DV. This latter line is a median of golden triangle BCD, hence passes through the spiral center Y, which therefore lies at the intersection of DV with CZ. Both constructions would involve drawing the perpendicular bisector of AB and a perpendicular to BC. One would use a circle, the other a straight line to intersect with CZ to locate Y. Although the straight line is more easily drawn than a circle, the circle method would give one a check on accuracy in passing through point D. In Figure 20-16 both constructions are shown; the redundancy confirms the accuracy of both approaches.

Q: *FIGURE 20-2 SHOWS GOLDEN TRIANGLE ABC SUBDIVIDED INTO THE SMALLER GOLDEN TRIANGLE BCD AND GOLDEN GNOMON BDA. USE THEOREM 2-1 TO COMPUTE THE RATIOS OF THE AREAS OF THE GOLDEN GNOMON AND THE TWO GOLDEN TRIANGLES.*

As the two golden triangles are geometrically similar, and triangle BCD has sides whose lengths are ϕ times as long as the corresponding ones in triangle ABC, the area of the smaller golden triangle is ϕ^2, or $(1-\phi)$ times that of the larger golden triangle. The area of the golden gnomon BDA is therefore ϕ times that of the larger golden triangle ABC. The smaller golden triangle BCD has an area ϕ times that of the golden gnomon BDA.

Notes

[1] Penrose, Roger: *The Role of Aesthetics in Pure and Applied Mathematics Research*, J. Inst. Mathematics and Its Applications, **10**, 266–271 (1974).

[2] Gardner, Martin: *Mathematical Games; Extraordinary Nonperiodic Tiling that Enriches the Theory of Tiles*, Scientific American, January, 1977.

[3] The remainder of this chapter is adapted from Loeb, A. L., and William Varney: *Is There a Golden Spiral, and If Not, Where Is Its Center?* in *Spiral Symmetry*, I. Hargittai, ed. World Science Books, Singapore, 47–61 (1991).

[4] Huntley, H. E.: *The Divine Proportion.* Dover, 1970.

XXI

Quasi Symmetry

In Chapter VI it was shown that a five-fold rotocenter in a plane precludes the existence of any other rotocenter in that plane. Since crystals necessarily have translational symmetry, it was believed that five-fold rotational symmetry cannot exist in crystals. However, recent experiments[1,2] produced x-ray diffraction patterns having five-fold rotational symmetry, a result which had previously been believed possible only if the crystals through which the x-ray beams had been diffracted were themselves five-fold symmetrical.

This apparent contradiction has led to a re-examination of the general question of what kind of order will produce diffraction patterns having discrete bright spots[3] and the conclusion that translational symmetry is not a necessary condition. As it happens, shortly beforehand, Roger Penrose[4] had investigated tilings of the plane having no translational symmetry. Penrose used two different types of tiles, one consisting of two golden triangles, the other of two golden gnomons. Although the original tiles were dart- and kite-shaped, they were later modified to rhombic forms, a slender one consisting of two golden triangles joined at their base and a fat one made up of two golden gnomons joined at their base (Figure 21-1).

Figure 21-1 *Penrose tiles: (a) golden triangle*

Figure 21-1 (b) *Golden gnomon*

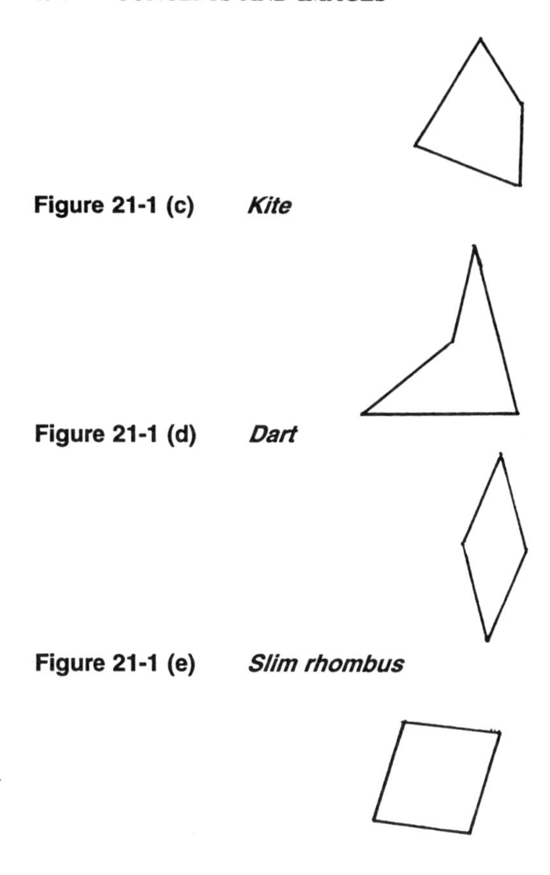

Figure 21-1 (c) ***Kite***

Figure 21-1 (d) ***Dart***

Figure 21-1 (e) ***Slim rhombus***

Figure 21-1 (f) ***Fat rhombus***

These rhombi can be joined according to certain matching rules[5,6] to form a pattern such as shown in in Figure 21-2. It will be observed that in this pattern there are five fat tiles meeting at a common vertex, but that this vertex is not a true five-fold rotocenter in the total context of the tessellation. Nevertheless, such patterns may produce sharp diffraction patterns[7]; although they do not have translational or rotational symmetry, they do possess some type of order, which is called *quasi-symmetry*.

De Bruijn (op. cit.), Paul Donchian and H. S. M. Coxeter[8] have found that quasi-symmetrical patterns may be obtained by projection of a higher-dimensional symmetrical pattern onto a lower dimension. Their findings do not imply that quasicrystals are three-dimensional manifestations of some higher dimensional minerals. Rather, they relate quasi-symmetry in a space of a given dimension to higher-dimensional symmetry. Quasi-symmetry describes an ordering system not previously understood, with the result that quasi-symmetrical patterns had not been recognized as orderly.

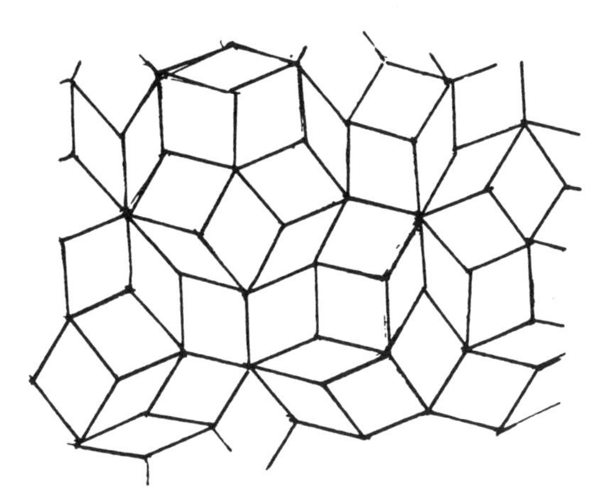

Figure 21-2 ***Portion of a Penrose tessellation***

We shall here discover a one-dimensional quasi-symmetrical pattern which actually turns out to be the projection of a selected portion of a two-dimensional lattice onto a straight line. First we shall generate some strings of symbols each of which may be either x or y.[9] Beginning with a single symbol, the principle for generating each string is that every x in the previous string is transformed into the pair of symbols x, y, every y into x:

$$T(x) = x\,y \qquad (21-1x)$$

$$T(y) = x \qquad (21-1y)$$

Thus, starting with x, the following strings are generated successively:

$$x$$
$$x\,y$$
$$x\,y\,x$$
$$x\,y\,x\,x\,y$$
$$x\,y\,x\,x\,y\,x\,y\,x$$
$$x\,y\,x\,x\,y\,x\,y\,x\,x\,y\,x\,x\,y$$
$$x\,y\,x\,x\,y\,x\,y\,x\,x\,y\,x\,x\,y\,x\,y\,x\,x\,y\,x\,y\,x$$

Q: *GENERATE THE NEXT STRING YOURSELF. HOW LONG IS EACH STRING, i.e., HOW MANY ITEMS DOES EACH CONTAIN? HOW MANY SYMBOLS x AND HOW MANY SYMBOLS y DO YOU FIND IN EACH STRING?*

Although these strings are generated by a very definite rigorous rule, it is not easy to perceive order in them. Nevertheless, a pattern of order emerges as one examines the strings more closely. In the first place, the length of each string appears to equal a Fibonacci number. That this observation is generally true is proven as follows. If we call the number of times the symbol x occurs in the i-th string $n_i(x)$, the number of times that y occurs in the i-th string $n_i(y)$, and the total length of the i-th string N_i, then

$$N_i = n_i(x) + n_i(y) . \qquad (21-2)$$

Since every x in a given string is transformed into $x\,y$ in the next string and every y in that given string is transformed into x, every item in the given string contributes an x to the next string, so that

$$n_i(x) = N_{i-1} . \qquad (21-3)$$

On the other hand, all symbols y in a given string are generated out of an x in the previous string, so that

$$n_i(y) = n_{i-1}(x) . \qquad (21-4)$$

From equation (21-3) there follows:

$$n_{i-1}(x) = N_{i-2} . \qquad (21-5)$$

Substitute equation (21-5) into equation (21-4):

$$n_i(y) = N_{i-2} . \qquad (21-6)$$

Finally, when equations (21-3) and (21-6) are substituted into equation (21-2):

$$N_i = N_{i-1} + N_{i-2} . \qquad (21-7)$$

This last equation is recognized as the recursion relation between successive Fibonacci numbers (cf. equation 19-3): Each number equals the sum of the two preceding numbers in the sequence. Since the length of the first string, 1, is a Fibonacci number, the lengths of all succeeding strings must be Fibonacci numbers.

From equations (21-5) and (21-6) it then follows that the individual numbers of symbols x and y are also Fibonacci numbers. Thus we have proven

Theorem 21-1: The length of each string generated by successive applications of T on x equals a Fibonacci number, as do the numbers of symbols x and y in any one string.

Q: *COMPARE A GIVEN STRING ITEM BY ITEM WITH THE ITEMS OF THE PREVIOUS STRING, STARTING AT THE LEFT END. WHAT DO YOU OBSERVE? WHAT ABOUT THE ITEMS STARTING AT THE RIGHT END?*

Let us denote the sequence of symbols in a given string by the notation S_i, that of the previous string by S_{i-1}. A string constituted of the items of S_i followed by those of S_{i-1} will be denoted S_iS_{i-1}. Then we observe that

Theorem 21-2: Any string S_i is constituted of the elements of the previous string S_{i-1} followed by the elements of the string preceding the latter, S_{i-2}.

Hence,

$$S_{i+1} = S_iS_{i-1}\,. \tag{21-8}$$

Any string S_i may therefore be written

$$S_i = X_iY_i\,, \tag{21-9}$$

$$\text{where}\quad X_i = S_{i-1} \tag{21-9x}$$

$$\text{and}\quad Y_i = S_{i-2} \tag{21-9y}$$

Remembering that the operation T transforms each string into the subsequent one, we find

$$TX_i = TS_{i-1} = S_i = X_iY_i\,, \tag{21-10x}$$

$$TY_i = TS_{i-2} = S_{i-1} = X_i\,. \tag{21-10y}$$

When equations (21-10x) and (21-10y) are compared with equations (21-1x) and (21-1y), we observe:

Theorem 21-3: Any string S_i may be decomposed into two *component* strings X_i and Y_i, each of which obeys the same transformation rules as did the original x and y.

The original x and y with which we had started may therefore themselves have been strings generated from even more elementary units. The transformation T has thus been shown to generate a highly ordered hierarchical structure whose pattern of order does not appear to be related to symmetry at all, but does relate to Fibonacci numbers. However, when we represent the strings generated from T repeatedly operating on x graphically, we shall discover such a relationship.

To this purpose, consider the lattice of Figure 21-3, a lattice in which every lattice point has four neighbors at the corner of a square. One of these lattice points is chosen as the origin, two mutually perpendicular axes, X and Y, join it to its neighbors. On this lattice we shall represent every string S_i as follows: The symbol x is represented by a step to the next lattice point to the right parallel to the X-axis, every y by a step to the next lattice point upward, parallel to the Y-axis; every string is represented by a path constituted of consecutive line segments parallel to one of the axes.

Figure 21-3 ***Square lattice***

Q: *ON A PIECE OF GRAPH PAPER PLOT THE FIRST EIGHT STRINGS GENERATED ABOVE.*

In Figure 21-4 we show the eigth string, which, of course, contains all previous component strings. We tabulate the ends of the component strings in terms of the number of steps in the X-direction, and those in the Y-direction as follows:

x	ends at 1, 0
$x\,y$	ends at 1, 1
$x\,y\,x$	ends at 2, 1
$x\,y\,x\,x\,y$	ends at 3, 2
$x\,y\,x\,x\,y\,x\,y\,x$	ends at 5, 3
$x\,y\,x\,x\,y\,x\,y\,x\,x\,y\,x\,x\,y$	ends at 8, 5
$x\,y\,x\,x\,y\,x\,y\,x\,x\,y\,x\,x\,y\,x\,y\,x\,x\,y\,x\,y\,x$	ends at 13, 8

Note that, as expected from Theorem 21-1, the numbers in these two columns equal Fibonacci numbers. Therefore the slope of the line joining the end of each successive string to the origin has the general formula (a_{n-1}/a_n). Recalling that

$$\phi^n = (-1)^{n+1}(a_n\phi - a_{n-1}), \tag{19-7}$$

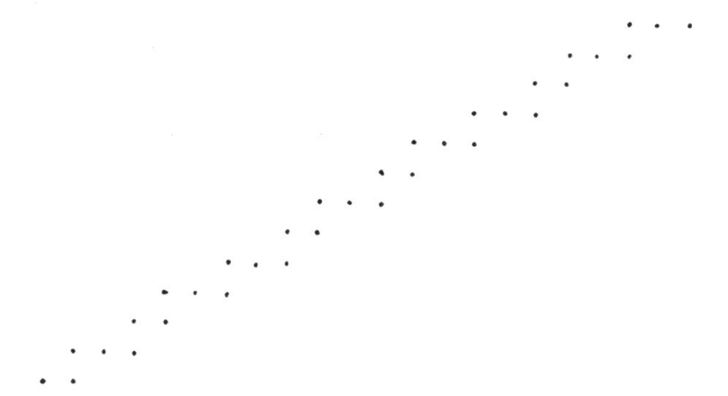

Figure 21-4 ***Lattice points defined by the string xyxxyxyxxyxxyxyxxyxyxxyxxyxyxxyxxy***

and that ϕ^n approaches zero as n increases, we recognize that the ends of successive strings will approach, alternately from above and below, but never quite reach, a line through the origin having slope ϕ. This latter line, after passing through the origin, will never pass through another lattice point, for if it did, then it would have a rational slope, and the slope ϕ is irrational. The equation of this line is

$$y = \phi x\,. \qquad (21-11)$$

This line is shown in Figure 21-5: It is the middle one of three middle mutually parallel ones drawn in that figure. The other two lines will be

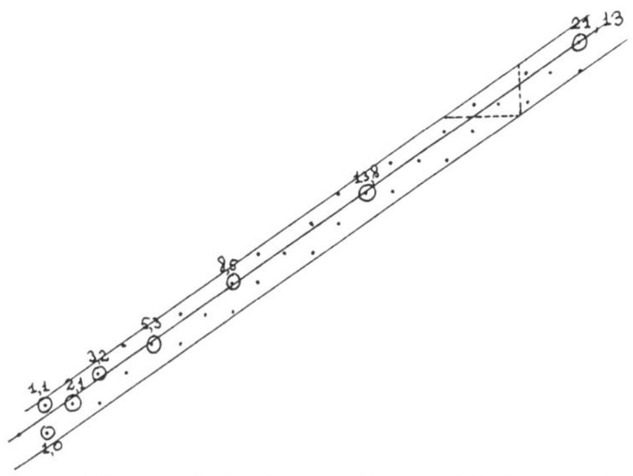

Figure 21-5 ***Lattice points from Figure 21-4, together with the ends of each string. Three lines are shown whose slope equals the golden fraction ϕ. Horizontal and vertical dotted lines inndicate the horizontal and vertical widths of the strip***

explained presently. Each string is represented by a path in which either one or two horizontal treads are followed by a single riser, never straying far from the line given by equation (21-11). We have explicitly labeled the pair of numbers chracterizing the end of each component string in Figure 21-5. The path passes through all lattice points lying within a strip bounded by two straight lines also having a slope ϕ, whose equations are, respectively,

$$y = \phi x + \phi \qquad (21-12)$$

and

$$y = \phi x - 1 \,. \qquad (21-13)$$

Each of these two lines passes through but a single lattice point, the former through the point $(-1,\ 0)$, the latter through $(0,\ -1)$. All three lines are shown in Figure 21-5.

Q: *ASCERTAIN WHETHER ALL POINTS REPRESENTING STRINGS WHICH YOU HAD PREVIOUSLY PLOTTED ACTUALLY LIE BETWEEN THESE LINES. EXTEND THE THREE SLOPING LINES A BIT TO THE LOWER LEFT, AND DRAW A SQUARE HAVING THREE VERTICES LOCATED AT THE LATTICE POINTS (0.0), (−1, 0), AND THE ORIGIN (0,0). WHAT DO YOU OBSERVE?*

The points $(-1,\ 0)$ and $(0,\ -1)$ lie at diagonally opposite corners of a square whose sides are one step long; the origin (0,0) is another vertex of this square. The two lines are exactly far enough apart to accommodate this small square. From equations (21-12) and (21-13) we can also find the lengths of line segments of any horizontal and any vertical line lying inside the strip delineated by the two sloping lines. A horizontal line $y = H$ will intersect the borders of the strip at $x = (H - \phi)/\phi$ and $x = (H + 1)/\phi$ respectively, so that a portion of length $x = (1 + \phi)/\phi$ lies within the strip. We shall call this distance the ***horizontal width*** of the strip. By an analogous argument we find the ***vertical width*** of the strip to be $(1 + \phi)$. Recalling that

$$(1+\phi)/\phi \;=\; (1+\phi)^2 \;=\; 1+2\phi+\phi^2 \;=\; 2+\phi\,,$$

we note that the horizontal width of the strip is $(2+\phi)$, the vertical width $(1+\phi)$.

Contemplating the path representing the strings generated by repeated operation of T on x, we note that it never exceeds the vertcal width of the strip, for its vertical risers are at most one step high, hence never exceed $(1 + \phi)$, and its horizontal treads are either one or two steps wide, hence never exceed the horizontal width $(2 + \phi)$. When the path turns 90°, it does so because it would have left the strip if it were to have continued to go straight ahead. The paths representing the strings pass through every lattice point lying within the strip bounded by the lines given by equations (21-12) and (21-13).

If we project all the lattice points within the strip just defined onto any line having slope ϕ, we obtain a one-dimensional array of points whose arrangement is orderly, though not what we have so far considered symmetrical. It is this observation, namely that:

1. A selected region of a highly symmetrical array in a higher dimension, when projected onto an "irrational" subspace of lower dimension, will generate an apparently unsymmetrical array; and
2. Such a projected array may produce a diffraction pattern of sharply defined points;

that has led to the naming of this apparently unsymmetrical projected array as "quasi-symmetrical."

We shall now project the lattice points inside the defined strip onto a line having slope ϕ. As shown in Figure 21-6, we can construct a golden rectangle having a diagonal in the X-direction (to be called an *X-diagonal*) and one having a diagonal in the Y-direction (a *Y-diagonal*). The two rectangles, having diagonals of identical lengths, are mutually congruent and oriented perpendicularly to each other, each with its sides parallel and perpendicular to the sloping line.

The projection of the X-diagonal onto the sloping line corresponds to the long side of these rectangles, whereas the projection of the Y-diagonal

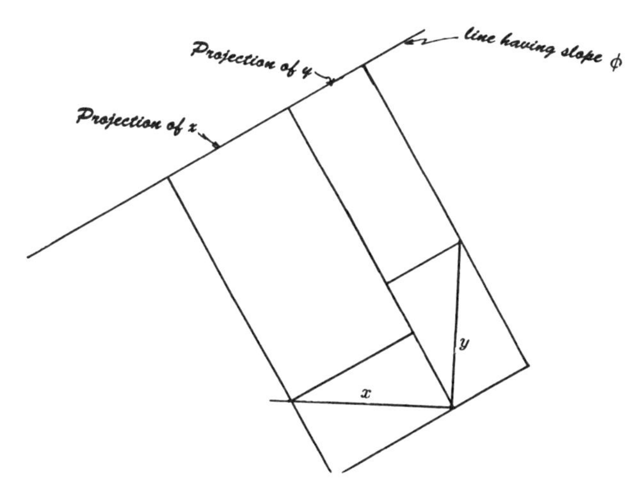

Figure 21-6 ***Projection of lattice points on a line having slope equal to the golden fraction***

onto the sloping line corresponds to the short side. Accordingly, the projection of a diagonal in the Y-direction is ϕ times as long as the projection of a diagonal in the X-direction. We shall accordingly define a coordinate along the sloping line such that the symbol x is represented along the sloping line by the distance ϕ, the symbol y by the distance ϕ^2. Since $\phi^2 = 1 - \phi$, the sequence xy will correspond to a unit distance in this coordinate along the sloping line.

When the graph of the string $xyxxyxyxxyxxyxyxxyxyx$ in Figure 21-5 is projected onto one of the lines having slope ϕ, the coordinates of the projections of its steps may then be tabulated as follows (Table 21-1):

Table 21-1: ***Projection of the Lattice Points Within the Strip Onto the Sloping Line.***

Path Element	*Projected Length*	*Distance from Origin*
x	$\boldsymbol{\phi}$	$\boldsymbol{\phi}$
y	**1-$\boldsymbol{\phi}$**	**1**
x	$\boldsymbol{\phi}$	**1+$\boldsymbol{\phi}$**
x	ϕ	$1+2\phi$
y	**1-$\boldsymbol{\phi}$**	**2 +$\boldsymbol{\phi}$**
x	ϕ	$2+2\phi$
y	$1-\phi$	$3+\phi$
x	$\boldsymbol{\phi}$	**3+2$\boldsymbol{\phi}$**
x	ϕ	$3+3\phi$
y	$1-\phi$	$4+2\phi$
x	ϕ	$4+3\phi$
x	ϕ	$4+4\phi$
y	**1-$\boldsymbol{\phi}$**	**5+3$\boldsymbol{\phi}$**
x	ϕ	$5+4\phi$
y	$1-\phi$	$6+3\phi$
x	ϕ	$6+4\phi$
x	ϕ	$6+5\phi$
y	$1-\phi$	$7+4\phi$
x	ϕ	$7+5\phi$
y	$1-\phi$	$8+4\phi$
x	$\boldsymbol{\phi}$	**8+5$\boldsymbol{\phi}$**
⋮	⋮	⋮

The terminals of component strings are shown in bold face: Not too surprisingly, their projections have Fibonacci numbers as coefficients.

Q: *PLOT THE LINEAR ARRAY OF POINTS TABULATED IN TABLE 21-1, BUT INSTEAD OF USING A VEY GOOD APPROXIMATION FOR THE VALUE OF ϕ, MAKE A SERIES OF PLOTS USING SUCCESSIVE APPROXIMATIONS FOR ϕ, STARTING WITH 1/2, THEN 2/3, NEXT 3/5, etc.*

We observe that the approximation $\phi = 1/2$ yields a linear lattice of equally spaced points, hence perfectly symmetrical. With increasingly good approximations, however, the locations of the points fluctuate around these equally spaced locations, illustrating the quasi-symmetry of the array.

Although these quasi-symmetrical arrays may lead to further understanding of packing and stacking arrangements in crystals, that in itself would not be sufficient reason to include them in the present book. What is important is that beyond applications to physical science, quasi-symmetry offers the artist, architect, designer and composer new principles of organization.

Q: *USING THE TRANSFORMATION*

$$T(x) = xy;\, T(y) = z;\, T(z) = xy\,,$$

GENERATE A SET OF STRINGS. HOW LONG IS EACH STRING? HOW MANY ELEMENTS X, Y AND Z OCCUR IN EACH STRING? CAN YOU PROVE A GENERAL EXPRESSION?

The transformation

$$T(x) = xy;\, T(y) = z;\, T(z) = xu;\, T(u) = z$$

generates the following strings:

$$x$$
$$x\,y$$
$$x\,y\,z$$
$$x\,y\,z\,x\,u$$
$$x\,y\,z\,x\,y\,x\,y\,z$$
$$x\,y\,z\,x\,u\,x\,y\,z\,x\,y\,z\,x\,u$$
$$x\,y\,z\,x\,u\,x\,y\,z\,x\,y\,z\,x\,u\,x\,y\,z\,x\,u\,x\,y\,z$$

Note that in this instance the length of each string again equals a Fibonacci number, as do the numbers of each symbol in each string. Also, once each symbol has been introduced (in the fourth string), each string consists of the previous string followed by the string preceding that last one. The number of symbols in string S_i, N_i is

$$N_i = n_i\,(x) + n_i(y) + n_i(z) + n_i(u)\,.$$

Since every x is produced by either x or z in the previous string,

$$n_i(x) = n_{i-1}(x) + n_{i-1}(z)\,.$$

Similarly,

$$\begin{aligned} n_i(y) &= n_{i-1}(x) \\ n_i(z) &= n_{i-1}(y) + n_{i-1}(u) \\ n_i(u) &= n_{i-1}(z)\,. \end{aligned}$$

Therefore,

$$\begin{aligned} N_i &= n_{i-1}(x)n_{i-1}(z) \\ &\quad + n_{i-1}(x) \\ &\quad + n_{i-1}(y) + n_{i-1}(u) \\ &\quad + n_{i-1}(z) \\ &= N_{i-1} + n_{i-2}(y) + n_{i-2}(u) + n_{i-2}(x) + n_{i-2}(z) \\ &= N_{i-1} + N_{i-2}\,. \end{aligned}$$

Accordingly, the recursion property of the Fibonacci series is once more satisfied, so that these numbers here also occur in the lengths of the strings.

NOTES

[1] Henley, Christopher L.: *Quasicrystal Order, Its Origins and Its Consequences: A Survey of Current Models*, Comments on Condensed Matter Physics, 58–117 (1987).

[2] Shechtman, D. I.: Blech, D. Gratias and J. W. Cahn: *A Metallic p-Phase with a Long-Ranged Orientational Order and No Translational Symmetry,* Phys. Rev. Lett., **53**, 1951 (1984).

[3] Senechal, Marjorie, and Jean Taylor: *Quasicrystals: The View from Les Houches*, The Mathematical Intelligencer, **12**, 54–64 (1990).

[4] Penrose, Roger: *The Role of Aesthetics in Pure and Applied Mathematics Research*, J. Inst. Mathematics and Its Applications, **10**, 266–271 (1974).

Gardner, Martin: *Mathematical Games: Extraordinary Nonperiodic Tiling that Enriches the Theory of Tiles*, Scientific American, **236**, 110–121 (1977).

[5] de Bruijn N. G.: *Algebraic Theory of Penrose's Nonperiodic Tilings of the Plane.* Kon. Nederlands Akad. Wetensch. Proc. Ser. A 84 (Indadationes Mathematicae 43), 38–66 (1981).

de Bruijn N. G.: *Quasicrystals and Their Fourier Transform.* Kon. Nederl. Akad. Wetensch. Proc. Ser A89 (Indagationes Mathematicae 48), 123–152 (1986).

de Bruijn N. G.: *Modulated Quasicrystals.* Kon. Nedrl. Akad. Wetensch. proc. Ser A90 (Indagationes Mathematicae 49), 121–132 (1987).

[6] Katz, A., and M. Duneau: *Quasiperiodic Crystals and Icosahedral Symmetry*, Journal de Physique, **47**, 181–196 (1981).

[7] Mackay, A.: *Crystallography and the Penrose Pattern*, Physica, 114A, 609–613 (1982).

[8] Coxeter, H. S. M.: *Convex Polytopes.* Dover, New York, 1973.

[9] What follows was inspired by the article by Senechal and Taylor (*op. cit*), but the author must take full responsibility for modifying their notation and approach in the present context.

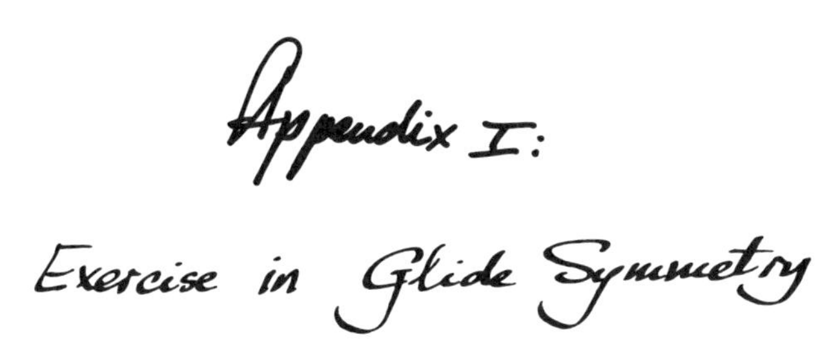

In Figure 7-12, reproduced here as Figure Appendix 1, there are twelve patterns. The reader was invited to discover which of these patterns have glide symmetry. Below we discuss the symmetry of each pattern.

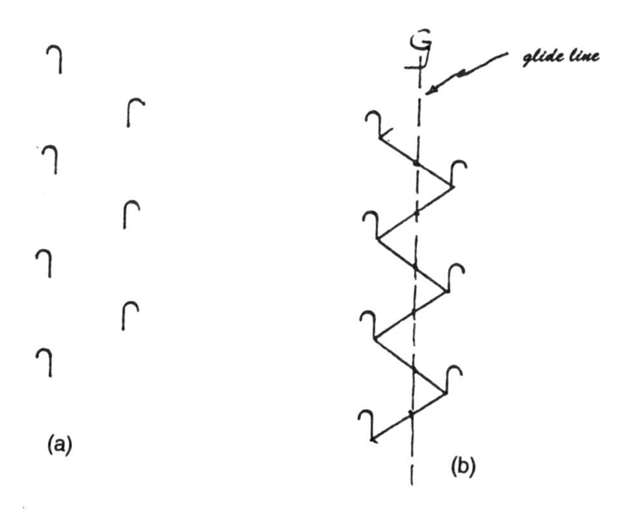

Figure Appendix I ***Find which patterns have, and which do not have, glide symmetry. Indicate why you think so.***

a) There is a horizontal mirror line: no glide symmetry.
b) The half-arrow motifs are all directly congruent: There is no enantiomorphy at all.
c) and e) In both, the motifs are enantiomorphically paired. However, in c) the spacing is such that each half-arrow is equidistant from both of its nearest neighbors, whereas in e) the motifs are clustered in pairs.

Therefore in e) the context of half the motifs is different from that of its enantiomorphs, with the result that e) has no glide symmetry, while c) does.

d) There is no enantiomorphy, hence no glide symmetry.

f) There is a glide line running perpendicular to the stems of the arrows.

g) There is no enantiomorphy, but note that there is two-fold rotational symmetry: The symmetry of this pattern is $22'\infty$.

h) There are glide lines parallel to and running between the stems.

Q: *PLACE A PIECE OF TRACING PAPER OVER PATTERN h), AND DRAW THE GLIDE LINES.*

i) There are two mutually perpendicular mirror lines in this pattern.

Q: *DRAW THESE LINES ON TRACING PAPER, AND LOCATE A ROTO-CENTER*

j) Although there are enantiomorphically paired motifs here, they are spaced such that the pattern has no enantiomorphy.

k) No enantiomorphy. The symmetry is actually the same as in g).

l) This one is tricky. The placement of the motifs would lead one to look for horizontal and vertical mirrors. However, such mirrors would reflect the entire array, so that additional strings of motifs would be generated. Since these are not present, there is no mirror symmetry. One might look for a glide line running diagonally from upper left to lower right. Reflected in such a glide line, the motifs would have enantiomorphs whose stems would not be vertical. Since such motifs are not present, the pattern has no glide symmetry. The symmetry of l) is accordingly $22'\infty$, like that of g) and k).

Appendix II:

Construction of Logarithmic Spiral

To construct a logarithmic spiral through the vertices of nested rectangles related to each other by dynamic symmetry (Figure 19-5), it is convenient to use two graphs, as illustrated in Figures Appendix 2 (a) and 2 (b). Along the horizontal axis of Figure Appendix 2 (a), locate a point A′ at a distance from the origin equal to the length of the line OA in the actual rectangle. The vertical axis is used to scale all radial distances relative to OA. On this vertical axis choose a point U at the top of the scale, assigning the distance of this point from the origin the value unity. Draw a straight line A′U. Place points B′, C′ and E′ on the horizontal axis, whose respective distances from the origin are the lengths of the line segments OB, OC and OE. Draw straight lines through B′, C′ and E′ parallel to A′U. Call the respective intercepts of these lines with the vertical axis B″, C″ and E″.

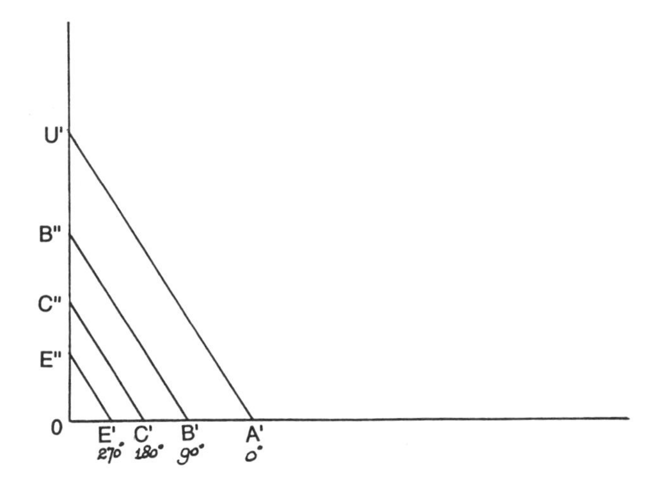

Figure Appendix 2 (a) ***Scaling the distances of A, B, C and D from O relative to OA***

Figure Appendix 2 (b) is drawn on semi-log paper; the horizontal axis represents the angular coordinate of the spiral, the vertical one the radial coordinate scaled relative to OA as unit distance. The points A, B, C and E are plotted on the semi-log paper: Their respective angular coordinates are 0°, 90°, 180° and 270°. Their scaled radial coordinates are unity for point A, and the respective distances from the origin of the points B″, C″ and E″ in Figure Appendix 2 (a). On this semi-log paper the points A, B, C and E should lie on a straight line, which should now be drawn.

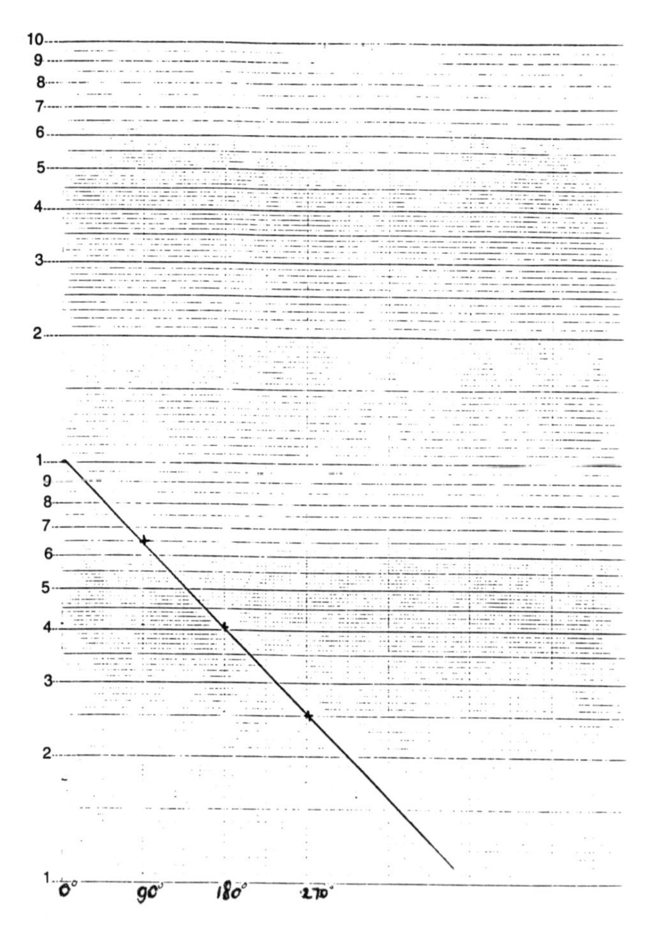

Figure Appendix 2 (b) ***Dynamic symmetry on semi-log paper***

The values of the coordinates of intermediate points on the spiral may now be read off Figure Appendix 2 (b): Intervals of 15° are recommended for the angular coordinate. The radial coordinates read off Figure Appendix

2 (b) are still scaled relative to OA as unity. When these scaled radial coordinates are recorded on the vertical axis of Figure Appendix 2 (a), and straight lines are drawn through these vertical intercepts parallel to the line UA′, these lines will intersect the horizontal axis at points whose distances from the origin just represent the radial distance from the center of the spiral. When all points at intervals of, say 15° are plotted, a French curve may be used to draw a smooth curve through them, which will represent the logarithmic spiral within the desired level of resolution.

A bibliography in Design Science encompasses many disciplines. William L. Hall, Jack C. Gray and the author have maintained a reading shelf and a file of reprints, which have been the major source of this appended bibliography on design science. As this bibliography transcends the boundaries of many disciplines, it has not been an easy task to keep abreast of developments. It has been the author's experience that a well-informed grapevine can be more effective than a computer search, and therefore he wishes to express his thanks to the many friends who sent in reprints and references. However, the author wishes to apologize for errors of commission and of omission which appear to be virtually inevitable.

BOOKS

Abbott, A: *Flatland*, Dover, New York (1952, 1983).

Albairn, K., J. Miall Smith, S. Steele and D. Walker: *The Language of Pattern*, Thames & Hudson, London (1974).

Abelson, H., and A. diSessa: *Turtle Geometry*, MIT press, Cambridge, MA (1981).

Adams, James L.: *Conceptual Blockbusting*, W. H. Freeman & Co., New York (1974).

Albers, Joseph: *Despite Straight Lines*, MIT Press, Cambridge, MA (1977).

Alexander, Christopher: *Notes on the Synthesis of Form*, Harvard University Press, Cambridge, MA (1964).

Alexander, Christopher, *et al.*: *A Pattern Language*, Oxford University Press, New York (1977).

Applewhite, E. J.: *Cosmic Fishing*, MacMillan, New York (1977).

Applewhite, E. J., ed.: *Synergetics Dictionary: The Mind of Buckminster Fuller*, Garland, New York (1986).

Arnheim, Rudolf: *Visual Thinking*, Univ. of California Press, Berkeley (1969).

Arnstein, Bennett: *Origami Polyhedra*, Exposition Press, New York (1968).
Audsley, W., and G. Audsley: *Designs and Patterns from Historic Ornament*, Dover, New York (1968, after the 1882 original).
Baglivo, Jenny, and Jack A. Graver: *Incidence and Symmetry in Design and Architecture*, Cambridge Univ. Press, Cambridge, U. K. (1983).
Ball, W. W. R., and H. S. M. Coxeter: *Mathematical Recreations and Essays*, Dover, New York (1986).
Banchoff, Thomas, and John Weber: *Linear Algebra through Geometry*, Springer, New York (1983).
Barratt, Krome: *Logic and Design, the Syntax of Art, Science and Mathematics*, Eastview, Westfield, NJ (1980).
Beard, Col. R. S.: *Patterns in Space*, Creative Publications, Palo Alto (1973).
Beauclair, B: *Art Nouveau, Patterns and Designs*, Bracken, London (1988).
Behnke, H., F. Bachman, K. Fladt and H. Kunle, eds.; S. H. Gould, transl.: *Fundamentals of Mathematics*, Volume II, MIT Press, Cambridge, MA (1974).
Belmont, Joseph: *L'Architecture création collective*, Les Editions Ouvrières, Paris (1970).
Belov, N. V.: *Derivation of the 230 Space Groups,* Leeds Philosophical and Literary Society, Leeds (1957).
Blackwell, William: *Geometry in Architecture*, Wiley, New York (1984).
Boles, Martha, and Rochelle Newman: *The Golden Relationship: Art, Math, Nature, I*, Pythagorean Press, Bradford, MA (1983).
Bool, Flip, ed.: *M. C. Escher (1898–1972)*, Haags Gemeentemuseun (1986).
Bourgoin, J.: *Arabic Geometrical Pattern and Design*, Dover, New York (1973).
Bragdon, Claude: *Projective Ornament*, Omen Press.
Bragg, Sir Lawrence: *Crystal Structures of Minerals*, Cornell Univ. Press, Ithaca, NY (1965).
Bragg, W. L.: *The Crystalline State*, MacMillan, New York (1934).
Bravais, M. A.; Amos J. Shalar, transl.: *On the Systems Formed by Points Regularly Distributed on a Plane or in Space*, Crystallographic Society of America (1949).
Brisson, David, ed.: *Hypergraphics*, AAAS Selected Symposium Series, Westview Press, Boulder (1978).
Buerger, Martin: *Elementary Crystallography*, Wiley, New York (1963).
Buerger, Martin: *Introduction to Crystal Geometry*, McGraw-Hill, New York (1971).
Burckhardt, J. J.: *Die Bewegungsgruppen der Kristallographie*, Birkhäuser, Basel (1966).

Burns, Gerald, and A. M. Glazer: *Space Groups for Solid State Scientists*, Academic Press, New York (1978).

Coffin, Stewart: *The Puzzling World of Polyhedral Dissections*, Oxford Univ. Press, New York (1990).

Collier, Graham: *Form, Space and Vision*, Prentice Hall, Englewood Cliffs, NJ (1963).

Cotton, F. Albert: *Chemical Applications of Group Theory*, Wiley, New York (1963).

Coxeter, H. S. M.: *Introduction to Geometry*, Wiley, New York (1961).

Coxeter, H. S. M., and W. O. Moser: *Generators and Relations for Discrete Groups*, Springer, New York (1965).

Coxeter, H. S. M., and S. L. Greitzer: *Geometry Revisited*, Mathematical Association of America, Washington, DC (1967).

Coxeter, H. S. M.: *Twelve Geometric Exercises*, Southern Illinois Univ. Press, Carbondale (1968).

Coxeter, H. S. M.: *Regular Polytopes*, Methuen, London (1948), Dover, New York (1973).

Coxeter, H. S. M.: *Complex Polytopes*, Cambridge Univ. Press, Cambridge, U. K. (1975).

Coxeter, H. S. M., M. Emmer, R. Penrose and M. L. Teuber, eds: *M. C. Escher, Art and Sciences*, Elsevier Science Publishers, Amsterdam (1986).

Critchlow, Keith: *Order in Space, A Design Source Book*, Voking Press, New York (1970).

Critchlow, Keith: *Islamic Patterns*, Schocken, New York (1976).

Dale, J. E., and F. L. Milthorpe, eds.: *The Growth and Functioning of Leaves*, Cambridge Univ. Press, New York (1982).

Dimond, S. J., and D. A. Blizard, eds.: *Evolution and Lateralization of the Brain*, New York Academy of Sciences (1977).

Doczi, Gyorgy: *The Power of Limits*, Shambala, Boulder (1981).

Doeringer, Suzannah, *et al.*, eds.: *Art and Technology*, MIT Press, Cambridge, MA (1970).

Doxiadis, C. A.: *Architectural Space in Ancient Greece*, MIT Press, Cambridge, MA (1972).

Drabkin, David: *Nature's Architecture*, Univ. of Pennsylvania Press, Philadelphia (1975).

Drew, Philip: *Frei Otto-Form and Structure*, Westview Press, Boulder (1976).

Dye, Daniel Sheets: *Chinese Lattice Designs*, Dover, New York (1974).

Edgerton, S. Y., Jr.: *The Renaissance Rediscovery of Linear Perspective*,

Basic Books, NY (1975).
Edmondson, Amy: *A Fuller Explanation: The Synergetic Geometry of R. Buckminster Fuller*, Birkhäuser Design Science Collection, A. L. Loeb, ed., Birkhäuser, Boston (1986).
Edwards, Betty: *Drawing on the Right Side of the Brain*, J. P. Tarcher, Los Angeles (1979).
Edwards, Edward B.: *Patterns and Designs with Dynamic Symmetry*, Dover, New York (1967).
Eisenhart, L. P.: *A Treatise on the Differential Geometry of Curves and Surfaces*, Dover, New York (1960).
El-Said, Issam, and Ayse Parman: *Geometric Concepts in Islamic Art*, World of Islam, London (1976).
Emerton, Norma: *The Scientific Reinterpretation of Form*, Cornell Univ. Press, Ithaca, NY (1984).
Engel, Peter: *Geometric Crystallography*, Reidel, Boston (1986).
Engel, Peter: *Folding the Universe*, Vintage, New York (1988).

[Note: These latter two references are by two different, identically named authors. The former is Swiss, the latter American.]

Ernst, Bruno: *The Magic Mirror of M. C. Escher*, Random House, New York (1976).
Escher, M. C., Giftwraps, Abrams, New York (1987).
Escher, M. C.: *The Graphic Work*, Meredith, New York (1967), and various other publishers.
Escher, M. C.: *Escher on Escher: Exploring the Infinite*, Abrams, New York (1989).
Falicov, L. M.: *Group Theory and Its Applications*, Univ. of Chicago Press, Chicago (1966).
Fedorov, E. S.; D. and K. Harker, transl.: *Symmetry of Crystals*, American Crystallographic Association (1971).
Fejes Toth, L.: *Regular Figures*, Macmillan, New York (1964).
Flegg, H. Graham: *From Geometry to Topology*, Crane Russak and Co., Inc., New York (1974).
Fleury, Michael: *Graphisme et géométrie*, Presses de l'Univ. de Québec, Québec (1986).
Fischer, W., H. Burzlaff, E. Hellner and J. D. H. Donnay: *Space Groups and Lattice Complexes*, U. S. Department of Commerce, National Bureau of Standards (1973).
Francis, George K.: *A Topological Picture Book*, Springer, New York (1987).
Fry, Charles R.: *Art Deco Designs in Color*, Dover, New York (1975).

Fuller, R. Buckminster: *No More Second-hand God*, S. Illinois Univ. Press, Carbondale (1963).

Fuller, R. Buckminster: *Intuition*, Doubleday, New York (1972).

Fuller, R. Buckminster: *Nine Chains to the Moon*, S. Illinois Univ. Press, Carbondale (1963).

Fuller, R. Buckminster: *Operating Manual for Spaceship Earth,* S. Illinois Univ. Press, Carbondale (1969).

Fuller, R. Buckminster, with E. J. Applewhite and A. L. Loeb: *Synergetics*, Macmillan, New York (1975).

Fuller, R. Buckminster, with E. J. Applewhite, ed.: *Synergetics 2*. Macmillan, New York (1979).

Fuller, R. Buckminster: *Critical Path*, St. Martin's Press, New York (1981).

Fuller, R. Buckminster: *Grunch of Giants*, St. Martin's Press, New York (1983).

Gabel, Medard: *Energy, Earth and Everyone,* Straight Arrow Books, San Francisco (1975).

Gardner, Martin: *Logic Machines and Diagrams*, McGraw-Hill, New York (1958).

Gardner, Martin: *Mathematical Puzzles and Diversions*, Simon & Schuster, New York (Vol. 1, 1959; Vol. 2, 1961).

Gardner, Martin: *The Ambidextrous Universe*, Scribner, New York (1979).

Gardner, Martin: *Penrose Tiles to Trapdoor Codes*, W. H. Freeman, New York (1988).

Gasson, Peter C.: *Geometry of Spatial Forms*, Wiley, New York (1983).

Ghyka, Matila: *The Geometry of Art and Life*, Sheed and Ward (1946), reprint Dover, New York (1977).

Ghyka, Matila: *Geometrical Composition and Design*, Alex Tiranti (1964).

Gillespie, Ronald J.: *Molecular Geometry*, Van Nostrand Reinhold, London (1972).

Gips, James: *Shape Grammars and Their Uses*, Birkhäuser, Boston (1975).

Gombrich, E. H.: *Art and Illusion*, Princeton Univ. Press, Princeton (1960).

Gombrich, E. H.: *The Sense of Order*, Cornell Univ. Press, Ithaca, NY (1984).

Gomez-Alberto, Perez: *Architecture and the Crisis of Modern Science,* MIT Press, Cambridge, MA (1984).

Gould, Stephen Jay: *Ontogeny and Philogeny,* Harvard Univ. Press, Cambridge, MA (1977).

Graziotti, Ugo Adriano: *Polyhedra: The Realm of Geometric Beauty,* San Francisco (1962).

Gregory, R. L.: *The Intelligent Eye*, McGraw-Hill, New York (1970).

Grenander, U.: *Pattern Synthesis*, Springer, New York (1976).
Grünbaum, Branko: *Convex Polytopes*, Interscience, New York (1967).
Grünbaum, Branko, and G. C. Shephard: *Tilings and Patterns*, W. H. Freeman, New York (1987).
Gullette, Margaret M.: *The Art of Teaching*, Alpine, Stoughton, MA (1982).
Hadanard, Jacques: *The Psychology of Invention in the Mathematical Field*, Dover, New York (1954).
Hahn, Theo, ed.: *International Tables for Crystallography*, Reidel, Boston, (1985, 1988).
Halmos, Paul R.: *Naive Set Theory*, Van Nostrand, New York (1960).
Harary, Frank, Robert Z. Norman, and Dorwin Cartwright: *Structural Models: An Introduction to the Theory of Directed Graphs*, Wiley, New York (1965).
Harary, Frank, ed.: *Topics in Graph Theory*, NY Acad. of Sciences, New York (1979).
Hargittai, István, ed.: *Symmetry, Unifying Human Understanding*, Pergamon, New York (Vol. 1, 1986; Vol. 2, 1989).
Hargittai, István, ed.: *Spiral Symmetry*, World Scientific, Singapore (1992).
Hargittai, István, ed.: *Five-fold Symmetry*, World Scientific, Singapore (1992).
Hargittai I, and M.: *Symmetrty through the Eye of a Chemist*, VCH Publishers, New York (1987).
Harlan, Calvin: *Vision and Invention*, Prentice Hall, Englewood Cliffs, NJ (1986).
Heesch, Heinrich: *Reguläres Parkettierungsproblem*, West Deutscher Verlag, Cologne (1968).
Heidegger, Martin: *On the Way to Language*, Harper and Row, New York (1971).
Henderson, Linda: *The Fourth Dimension and Non-Euclidean Geometry in Modern Art*, Princeton Univ. Press, Princeton (1983).
Hilbert, David, and S. Cohn-Vossen: *Geometry and the Imagination*, Chelsea, New York (1952).
Hildebrandt, Stefan, and Anthony Tromba: *Mathematics and Optimal Form*, W. H. Freeman, New York (1985).
Hill, Anthony: *DATA: Directions in Art, Theory and Aesthetics*, Farber and Farber, New York (1968).
Hilliard, Garland K., ed.: *Proceedings of the International Conference on Descriptive Geometry*, Wm. Brown, Dubuque, IA (1978).
Hilton, Harold: *Mathematical Crystallography*, Dover, New York (1963).
Hoag, John: *Islamic Architecture*, Rizzoli, New York (1987).
Hofstadter, Douglas R.: *Gödel, Escher, Bach*, Basic Books, New York (1979).
Holden, Alan, and Phylis Singer: *Crystals and Crystal Growing*, Anchor, Garden City, NJ (1960).

Holden, Alan: *The Nature of Solids*, Columbia Univ. Press, New York (1965).
Holden, Alan: *Shapes, Space and Symmetry*, Columbia Univ. Press, New York (1971).
Holden Alan: *Bonds Between Atoms*, Oxford Univ. Press, New York (1971).
Holden, Alan: *Orderly Tangles*, Columbia Univ. Press, New York (1983).
Huntley, H. E.: *The Divine Proportion*, Dover, New York (1970).
International Tables for X-ray Crystallography, Reidel, Dordrecht/Boston (1983 ff).
Ivins, William M., Jr.: *Art and Geometry*, Harvard Univ. Press, Cambridge (1946), Dover, New York (1969).
Jaswon, M. A.: *An Introduction to Mathematical Crystallography*, American Elsevier, New York (1965).
Jones, Owen: *The Grammar of Ornament*, Van Nostrand Reinhold, New York (1968).
Judson, Horace F.: *The Search for Solutions*, Holt, Rinehart & Winston, New York (1980).
Kappraff, Jay: *Connections, the Geometric Bridge between Art and Science*, McGraw-Hill, New York (1990).
Kenner, Hugh: *Bucky, A Guided Tour of Buckminster Fuller*, Wm. Morrow, New York (1973).
Kenner, Hugh: *Geodesic Math and How to Use It*, Univ. of California Press, Berkeley (1976).
Kepes, G., ed.: *The Module, Proportion, Symmetry and Rhythm*, Braziller, New York (1965).
Kepler, Johann: *The Six-Cornered Snowflake*, Oxford Univ. Press, London (1966).
Kim, Scott: *Inversions*, Byte Books/McGraw-Hill, New York (1981).
Kitaigorodskiy, A. I.: *Order and Disorder in the World of Atoms*, Springer, New York (1967).
Kline, Morris: *Mathematics in Western Culture*, Oxford Univ. Press, New York (1953).
Kline, Morris: *Mathematical Thought from Ancient to Modern Times*, Oxford Univ. Press, New York (1972).
Kline, Morris: *Mathematics and the Physical World*, Dover, New York (1981).
Lakatos, Imre: *Proofs and Refutations*, Cambridge Univ. Press, New York (1976).
Lauwerier, Hans: *Fractals* (in Dutch). Aramith, Amsterdam (1987).
Lauwerier, Hans: *Symmetrie-Regelmatique Strukturen in de Kunst* (in Dutch). Aramith, Amsterdam (1988).
Lawler, Robert: *Sacred Geometry: Philosophy and Practice*, Thames and Hudson, New York (1982).
Laycock, Mary: *Bucky for Beginners*, Activity Resources, Hayward, CA (1984).

LeCorbeiller, Philippe, ed.: *The Language of Science*, Basic Books, New York (1963).
LeCorbeiller, Philippe: *Dimensional Analysis*, Appleton Century Crofts, New York (1966).
Leeman, Fred: *Hidden Images*, Abrams, New York (1975).
Lehner, Ernst: *Alphabets and Ornaments*, Dover, New York (1968, after the original of 1952).
Lindgren, H.: *Geometric Dissections*, Dover, New York (1972).
Locher, J. L., ed.: *De Werelden van M. C. Escher*, Meulenhoff, Amsterdam (1971). Translated as *The World of M. C. Escher*, Abradale, Steven Sterk, Utrecht (1988).
Locher, J. L.: *Escher*, Thames and Hudson, London (1982).
Loeb, A. L.: *Color and Symmetry*, Wiley, New York (1971), Krieger, New York (1978).
Loeb, A. L.: *Crimson Heather, Twenty-one Scottish Country Dances*, Boston Branch Book Store, Royal Scottish Country Dance Society, Boston (1986).
Loeb, A. L.: *Space Structures, Their Harmony and Counterpoint*, Addison-Wesley, Reading, MA (1976), Fifth revised printing Birkhäuser Basel/Boston/Berlin (1991).
Lord, Athena: *Pilot for Spaceship Earth*, MacMillan, New York (1978).
MacGillavry, Carolina: *Fantasy and Symmetry*, Abrams, New York (1976).
Malina, Frank J., ed.: *Visual Art, Mathematics and Computers.* Pergamon, New York (1979).
Mandelbrot, Benoit B.: *The Fractal Geometry of Nature*, W. H. Freeman, New York (1983).
Maor, Eli: *To Infinity and Beyond*, Birkhäuser, Boston (1987).
March, Lionel, and Philip Stedman: *The Geometry of Environment, an Introduction to Spatial Organization in Design*, MIT Press, Cambridge, MA (1978).
Marks, R.: *The Dymaxion World of R. Buckminster Fuller*, Reinhold, Princeton (1960), Doubleday, New York (1973).
Martin, George E.: *Transformational Geometry: An Introduction to Symmetry*, Springer, New York (1982).
Martin, Kenneth: *Chance and Order*, Hillingdon Press, Uxbridge, UK (1973).
Martinez, Benjamin, and Jacqueline Block: *Visual Forces*, Prentice Hall, Englewood Cliffs, NJ (1988).
McHale, John R.: *Buckminster Fuller*, Braziller, New York (1962).
McMullen, P., and G. C. Shephard: *Convex Polytopes and the Upper Bound Conjecture*, Mathematics Society Lecture Note Series, no. 3, Cambridge Univ. Press, New York (1971).

McWeeny, R.: *Symmetry—An Introduction to Group Theory*, MacMillan, New York (1963).

Miller, Arthur: *Imagery in Scientific Thought*, Birkhäuser, Boston (1984).

Minsky, M., and S. Papert: *Perceptrons*, MIT Press, Cambridge, MA (1972, 1988).

Mitchell, William J.: *Computer Aided Design*, Van Nostrand Reinhold, New York (1977).

Miyazaki, Koji: *Form and Space, Polygons, Polyhedra and Polytopes*, Wiley, New York (1986).

Müller, Edith: *Gruppentheoretische und Strukturanalytische Untersuchungen der Maurischen Ornamente aus der Alhambra*, Rüschlikon, Switzerland (1944).

Nagel, Ernst, and James R. Newman: *Gödel's Proof*, New York Univ. Press, New York (1958).

Needham, Joseph: *Order and Life*, MIT Press, Cambridge, MA (1968).

Nerlick, Graham: *The Shape of Space*, Cambridge Univ. Press, New York (1976).

O'Daffer, P. G.: *Geometry: An Investigative Approach*, Addison Wesley, Reading, MA (1976).

Oman, Charles C., and Jean Hamilton: *Wallpapers*, Philip Wilson, London (1982).

Otto, Frei: *Architecture et bionique: Constructions naturelles.* Editions Delta and Spes, Denges, Switzerland (1985).

Pearce, Peter: *Structure in Nature Is a Strategy for Design*, MIT Press, Cambridge, MA (1978).

Pearce, Peter, and Susan Pearce: *Experiments in Form*, Van Nostrand Reinhold, New York (1980).

Pedersen, Jean, and A. Kent: *Geometric Playthings*, Price, Stern and Stone, Los Angeles.

Pedoe, Dan: *Geometry and the Visual Arts*, Dover, New York (1976).

Phillips, F. C.: *An Introduction to Crystallography*, Wiley, New York (1962).

Polya, G.: *How to Solve It*, Anchor Books, Garden City, NJ (1957).

Preparata, F. P., and M. I. Shamos: *Computational Geometry*, Springer, New York (1985).

Prince, E.: *Mathematical Techniques in Crystallography and Materials Science*, Springer, New York (1982).

Ranucci, E. R., and J. L. Teeters: *Creating Escher-type Drawings*, Creative Publications, Palo Alto, CA (1977).

Rawson, Philip: *Design*, Prentice Hall, Englewood Cliffs, NJ (1987).

Reichardt, Jasia: *Cybernetic Serendipity*, Studio International, New York (1968).

Rhodin, Johannes A. G.: *An Atlas of Ultrastructure*, W. B. Saunders, Philadelphia (1963).

Rosen, Joe: *Symmetry Discovered*, Cambridge Univ. Press, New York (1975).
Row, T. Sundra: *Geometric Exercises in Paper Folding*, Dover, New York (1987, reprint of 1905 original).
Rucker, Rudy: *Infinity and the Mind*, Birkhäuser, Boston (1982).
Ryschkewitsch, G.: *Chemical Bonding and the Geometry of Molecules*, Van Nostrand Reinhold, New York (1963).
Safdie, Moshe: *For Everyone a Garden*, MIT Press, Cambridge, MA (1974).
Sands, Donald E.: *Introduction to Crystallography*, W. A. Benjamin Advanced Book Program, New York (1969).
Saunders, Kenneth: *Hexagrams.* Parkwest, New York.
Schattschneider, Doris, and Walter Wallace: *M. C. Escher Kaleidocycles*, Ballantine, New York (1977).
Schattscneider, Doris: *Visions of Symmetry: Notebooks, Periodic Drawings and Related Work of M. C. Escher*, W. H. Freeman, New York (1990).
Schuh, Fred: *The Master Book of Mathematical Recreations*, Dover, New York (1968).
Schuyt, M. and Joost Elffers: *Anamorphoses*, Abrams, New York (1975).
Seiden, Lloyd: *Buckminster Fuller's Universe*, Plenum, New York (1989).
Segal, Gerry: *Synergy Curriculum*, Board of Education, New York (1979).
Senechal, Marjorie, and George Fleck, eds.: *Patterns of Symmetry,* Univ. of Massachusetts Press, Amherst (1977).
Senechal, Marjorie, and George Fleck, eds.: *Structure of Matter and Patterns of Science*, Schenkman, Cambridge, MA (1979).
Senechal, Marjorie, and George Fleck, eds.: *Shaping Space: A Polyhedral Approach*, Birkhäuser Design Science Collection, A. L. Loeb, ed. Birkhäuser, Boston (1988).
Shahn, Ben: *The Shape of Content*, Vintage Books, New York (1957).
Sheppard, Roger, Richard Threadgill and John Holmes: *Paper Houses*, Schocken, New York (1974).
Shubnikov, A. V., and B. A. Belov: *Colored Symmetry*, Macmillan, New York (1964).
Shubnikov, A. V., and B. A. Koptsik: *Symmetry in Science and Art*, Plenum, New York (1972).
Smith, Cyril S.: *From Art to Science*, MIT Press, Cambridge, MA (1980).
Smith, Cyril S.: *A Search for Structure*, MIT Press, Cambridge, MA (1980).
Snyder, Robert: *Buckminster Fuller, An Autobiographical Monologue/Scenario*, St. Martin's Press, New York (1980).
Steinhaus, H.: *Mathematical Snapshots*, Oxford Univ. Press, New York (1969).
Stevens, Garry: *The Reasoning Architect: Mathematics and Science in Design*, McGraw-Hill, New York (1990).
Stevens, Peter S.: *Patterns in Nature*, Little Brown, Boston (1973).

Stevens, Peter S.: *Handbook of Regular Patterns*, MIT Press, Cambridge, MA (1981).

Stonerod, David: *Puzzles in Space*, Stokes, Palo Alto, CA (1982).

Strache, Wolf: *Forms and Patterns in Nature*, Pantheon, New York (1956).

String Art Encyclopedia, Stirling, New York (1976).

Stuart, Duncan: *Polyhedral and Mosaic Transformations*, Univ. of North Carolina Press (1963).

Swirnoff, Lois: *Dimensional Color*, Birkhäuser Design Science Collection, A. L. Loeb, ed. Birkhäuser, Boston (1988).

Terpstra, P.: *Introduction to the Space Groups*, Univ. of Groningen Crystallographic Institute, Groningen, the Netherlands (1955).

Thom, Rene: *Stability and Morphogenesis*, Benjamin, Reading, MA (1975).

Thompson, d'Arcy Wentworth: *On Growth and Form*, Cambridge Univ. Press, London and New York (1942); abridged edition by J. T. Bonner, Cambridge Univ. Press (1969).

Thompson, James B.: *SMART-Readings in Science and Art*, Red Clay Books, Charlotte, NC (1975).

Tilings in the Collection of the Cooper Hewitt Museum, Smithsonian Institution, Washington, D. C. (1980).

Trudeau, Richard J.: *Dots and Lines*, Kent State Univ. Press, Kent, OH (1976).

Vasarely, Oeuvres Profondes Cinétiques, Ed. Griffon, Neuchatel (1973).

Verneuil, Ad., and M. P. Verneuil: *Abstract Art, Patterns and Designs*, Bracken, London (1988).

Vilenkin, N. Y.: *Stories about Sets*, Academic Press, New York (1968).

Vinci, Leonardo da; Pamela Taylor, ed.: *The Notebooks of Leonardi da Vinci*, Plume Books, New York (1960).

Walter, Marion: *The Mirror Puzzle Book*, Tarquin, Norfolk, UK (1985).

Washburn, Dorothy, and Donald W. Crowe: *Symmetries of Culture — Theory and Practice of Plane Pattern Analysis*, Univ. of Washington Pres, Seattle (1988).

Wechsler, Judith, ed.: *On Aesthetics in Science*, MIT Press, Cambridge, MA (1978), reprint Birkhäuser Design Science Collection, A. L. Loeb, ed., Birkhäuser, Boston (1988).

Weisberger, Edward, ed.: *The Spiritual in Art: Abstract Painting 1890–1985*, Los Angeles County Museum of Art, Los Angeles (1986).

Weyl, Hermann: *Symmetry*, Princeton Univ. Press, Princeton (1952).

Whyte, L. L., ed.: *Aspects of Form*, American Elsevier, New York (1968).

Whyte, L. L., A. G. Wilson and Donna Wilson, eds.: *Hierarchical Structures*, American Elsevier, New York (1969).

Wigner, Eugene P.: *Symmetries and Reflections*, Oxbow Press, Woodbridge, CT (1979).

Wilson, Forrest: *Architecture: A Book of Projects for Young Adults.* Van

Nostrand Reinhold, New York (1968).
Winter, John: *String Sculpture*, Creative Publications, Palo Alto, CA (1972).
Wittkower, Rudolf: *Idea and Image*, Thames and Hudson, London (1978).
Wolfe, Caleb Wroe: *Manual for Geometric Crystallography*, Edwards Bros., Ann Arbor, MI (1953).
Wong, Wucius: *Principles of Three-dimensional Design*, Van Nostrand Reinhold, New York (1968, 1977).
Yale, Paul: *Geometry and Symmetry*, Holden-Day, San Francisco (1968).
Young, Arthur M.: *The Reflexive Universe*, Delacorte, San Francisco (1976).
Zeier, Franz: *Paper Constructions*, Scribner, New York (1980).

ARTICLES

Baracs, Janos, ed.: *La revue topologie structurale*, published at irregular intervals, UQAM, C. P. 888, Succ. A, Montréal, Québec Canada.
Bolker, Ethan: *A Topological Proof of a Well Known Fact about Fibonacci Numbers*, The Fibonacci Quarterly, **15**, 245 (1977).
Bolker, Ethan: *Simplicial Geometry and Transportation Polytopes*, Trans. Amer. Math. Soc., **217**, 121–142 (1976).
Bombieri, E., and J. Taylor: *Quasicrystals, Tilings, and Algebraic Number Theory: Some Preliminary Connections*, Contemp. Math., **64**, 241–264 (1987).
Chorbachi, W. K., and A. L. Loeb: *An Islamic Pentagonal Seal from Scientific Manuscripts of the Geometry of Design*, in: *Fivefold Symmetry*, Istvan Hargittai, ed., 283–305, World Scientific, Singapore (1992).
de Bruijn, N. G.: *Algebraic Theory of Penrose's Non-periodic Tilings of the Plane*, Kon. Nederl. Akad. Wetensch. Proc. Ser. A 84 (Indignationes Mathematicae 43), 38–66 (1981).
de Bruijn, N. G.: *Quasicrystals and their Fourier Transform*, Kon. Nederl. Akad. Wetensch. Proc. Ser. A 89 (Indignationes Mathematicae 48), 123–152 (1986).
de Bruijn, N. G.: *Modulated Quasicrystals*, Kon. Nederl. Akad. Wetensch. Proc. Ser. A 90 (Indigationes Mathematicae 49), 121–132 (1987).
Cahn, J., and J. Taylor: *An Introduction to Quasicrystals*, Contemp. Math., **64**, 265–286 (1987).
Chorbachi, Wasma'a: *In the Tower of Babel: Beyond Symmetry in Islamic Design*, in *Symmetry, Unifying Human Understanding*, Istvan Hargittai, ed., Pergamon, New York (1989).
Chorbachi, Wasma'a, and A. L. Loeb: *A Pentagonal Seal*, in *Five-fold Symmetry*, Istvan Hargittai, ed., World Science Books, Singapore, 283–305 (1991).

Delone, B., N. Dolbilin, M. Shtogrin, and R. Galiulin: *A Local Criterion for the Regularity of a System of Points*, Soviet Math. Dokl. **17**, 319–322 (1976).

Erickson, R. O.: *Phyllotaxis and Shoot Form*, Botanical Soc. of Amer., Minneapolis, MN (1972).

Erickson, R. O.: *The Geometry of Phyllotaxis*, in *The Growth and Functioning of Leaves*, J. E. Dale and F. L. Milthorp, eds., Cambridge Univ. Press, New York (1982).

Erickson, R. O., and Alburt Rosenberg: *Phyllotaxis of the Sunflower Head*, Bot. Soc. Amer., Fort Collins, CO (1984).

Fejes Toth, L.: *What the Bees Know and What They Do Not Know*, Bull. Amer. Math. Soc., **20**, 468–481 (1964).

Gardner, Martin: *Mathematical Games: Extraordinary Nonperiodic Tiling that Enriches the Theory of Tiles*, Scientific American, January 1977.

Godrèche, C., and J. M. Luck: *Quasiperiodicity and Randomness in Tilings of the Plane*, J. Stat. Phys., **55**, 1–28 (1989).

Grünbaum, Branko: *Polytopes, Graphs, and Complexes*, Bull. Amer. Math. Soc., **76**, 1131–1201 (1970).

Grünbaum, Branko, and G. C. Shepard: *Satins and Twills: An Introduction to the Geometry of Fabrics*, Mathematics Magazine, **53**, 131–161 and 313 (1980).

Grünbaum, Branko, and G. C. Shepard: *A Hierarchy of Classification Methods for Patterns*, Zeitschrift f. Kristallographie, **154**, 163–187 (1981).

Haughton, Eric C., and A. L. Loeb: *Symmetry: A Case History of a Program*, J. Research in Science Teaching, **2**, 132ff (1964).

Henley, Christopher L.: *Quasicrystal Order, Its Origins and Its Consequences: A Survey of Current Models*, Comments on Condensed Matter Physics, 58–117 (1987).

Hoffer, William: *A Magic Ratio Recurs Throughout Art and Nature*, Orion Nature Quarterly, **4**, #1, 28–38 (1985).

Horgan, John: *Quasicrystals, Rules of the Game*, Science, **247**, March 2, 1990.

Katz A., and M. Duneau: *Quasiperiodic Patterns and Icosahedral Symmetry*, J. de Physique, **47**, 181–196 (1986).

Katz, A.: *Theory of Matching Rules for the 3-dimensional Penrose Tilings*, Commun. Math. Phys., **118**, 263–288 (1988).

Krull, Wolfgang; Betty S. Waterhouse, and William C. Waterhouse, transl.: *The Aesthetic Viewpoint in Mathematics*, The Math. Intelligencer, **9**, # 1, Springer, New York (1987).

La Brecque, M.: *Opening the Door to Forbidden Symmetries*, Mosaic (Nat'l Science Foundation), **18**, 2-23 (1987–88).

Levitov, L., and J. Rhyner: *Crystallography of Quasicrystals; Applications to Icosahedral Symmetry*, J. Phys. France, **49**, 1835–1849.

Loeb, A. L.: *Remarks on Some Elementary Volume Relations between Familiar Solids*, Mathematics Teacher, **50**, 417 (1965).

Loeb, A. L., and E. C. Haughton: *The Programmed Use of Physical Models*, J. Programmed Instruction, **3**, 9–18 (1965).

Loeb, A. L.: *The Architecture of Crystals*, in *Module, Proportion, Symmetry, Rhythm*, Gyorgy Kepes, ed., Vision and Value Series, Braziller, New York (1966).

Loeb, A. L.: *Structure and Patterns in Science and Art*, Leonardo, **4**, 339–345 (1971) reprinted in *Visual Art, Mathematics and Computers*, Frank J. Malina, ed., Pergamon Press (1979).

Loeb, A. L.: *Comments on David Drabkin's Book: Fundamental Structure: Nature's Architecture*, Leonardo, **10**, 313–314 (1977).

Loeb, A. L.: *Algorithms, Structure and Models*, in *Hypergraphics*, AAAS Selected Symposium Series, 49–68 (1978).

Loeb, A. L.: *A Studio for Spatial Order*, Proc. International Conf. on Descriptive Geometry and Engineering Graphics, Engineering Graphics Division, Amer. Soc. Engineering Education, 13-20 (1979).

Loeb, A. L.: *Natural Structure in the Man-made Environment*, The Environmentalist, **2**, 43–49 (1982).

Loeb, A. L.: *Escher, A Review of the book edited by J. L. Locher*, London Times Literary Supplement, September 17, 1982.

Loeb, A. L.: *On My Meetings and Correspondence between 1960 and 1971 with Graphic Artist M. C. Escher*, Leonardo, **15**, 23–27 (1982).

Loeb, A. L.: *Synergy, Sigmoids and the Seventh-Year Trifurcation*, The Environmentalist, **3**, 181–186 (1983) reprinted in Chrestologia, XIV #2, 4–8 (1989).

Loeb, A. L.: *Symmetry and Modularity*, J. Comp. & Math. with Appl. **128**, Nos. 1/2, 63–75 (1986), reprinted in *Symmetry, Unifying Human Understanding*, Istvan Hargittai, ed. Pergamon Press, NY, 63–75 (1986).

Loeb, A. L.: *Symmetry in Court and Country Dance*, J. Comp. & Math. with Appl., 629–639 (1986) reprinted as above, 629–639 (1986).

Loeb, A. L.: *Polyhedra in the Work of M. C. Escher*, in *M. C. Escher: Art and Science*, H. S. M Coxeter et al., ed. Elsevier Science Publishers, Amsterdam 195–202 (1986).

Loeb, A. L.: *Color and Symmetry: Ordering the Environment with Pattern*, Color and Light, **17**, 16–18 (1987).

Loeb, A. L.: *The Magic of the Pentangle: Dynamic Symmetry from Merlin to Penrose*, J. Comp. & Math. with Appl., **17**, 33–48 (1989).

Loeb, A. L.: *Hierarchical Structure and Pattern Recognition in Minerals and Alloys*, Per Mineralogia, **59**, 197–217 (1990).

Loeb, A. L.: *Reflections on Rotations Connections and Notations*, in *Symmetry: Culture and Science*, **1**, 39–56 (1990).

Loeb, A. L.: *Forces Shaping the Spaces We Inhabit*, International J. Space Structures, **6**, 281–286 (1991).

Loeb, A. L.: *On Behaviorism, Causality and Cybernetics*, Leonardo, **24**, 299–302 (1991).

Loeb, A. L.: *Can a Renaissance Person Survive in a Competitive Culture?*, International J. of Social Education, **1**, 24–40 (1992).

Loeb, A. L., and William Varney: *Does the Golden Spiral Exist, and If Not, Where Is Its Center?*, in *Spiral Symmetry*, Istvan Hargittai, ed. World Science Books, 47–61 (1991).

Mackay, A.: *Crystallography and the Penrose Pattern*, Physics, 114A, 609–613 (1982).

Masunaga, David: *Making Mathematics Appeal to the Mind and the Eye*, Harvard Grad. Sch. of Education Alumni Bull., Fall-Winter 1987, 18–21.

Nelson, David R.: *Quasicrystals*, Scientific American, **255**, 42 (1986).

Pawley, G. S.: *The 227 Tricontahedra*, **4**, # 221 (1975).

Penrose, Roger: *The Role of Aesthetics in Pure and Applied Mathematics Research*, J. Inst. Mathematics and Its Applications, **10**, 266–271 (1974).

Schattschneider, Doris: *The Plane Symmetry Groups: Their Recognition and Notation*, Amer. Math. Monthly, **85**, 441 (1978).

Schattschneider, Doris: *Tiling the Plane with Congruent Pentagons*, Math Magazine, **51**, 29–44 (1980).

Schulte, Egon: *Tiling Three-space by Combinatorially Equivalent Convex Polytopes*, Proc. London Math. Soc., **49**, 3, 128–140 (1984).

Senechal, Marjorie, and Jean Taylor: *Quasicrystals: The View from Les Houches*, The Mathematics Intelligencer, **12**, #2, 54–64 (1990).

Shechtman, D., I. Blech, D. Gratias and J. Cahn: *Metallic Phase with Long-range Orientational Order and No Translational Symmetry*, Phys. Rev. Letters, **53**, 1951–1954 (1984).

Smith, Cyril S.: *Grain Shapes and Other Metallurgical Applications of Topology*, Metal Interfaces, Amer. Soc. for Metals; University Microfilms, Ann Arbor, MI, OP 13754, 65–113.

Von Baeyer, Hans C.: *Impossible Crystals*, Discover, (February 1990.)

Index